Textbook of Geometry

Textbook of Geometry

April Watson

NY RESEARCH PRESS
New York

Published by NY Research Press
118-35 Queens Blvd., Suite 400,
Forest Hills, NY 11375, USA
www.nyresearchpress.com

Textbook of Geometry
April Watson

International Standard Book Number: 978-1-63238-826-1 (Hardback)

Cataloging-in-Publication Data

Textbook of geometry / April Watson.
p. cm.
Includes bibliographical references and index.
ISBN 978-1-63238-826-1
1. Geometry. 2. Mathematics. I. Watson, April.
QA445 .T49 2020
516--dc23

Printed in China.

Contents

Preface

The branch of mathematics which is concerned with questions of size, shape, relative position of objects and the properties of space is known as geometry. Some of the important concepts in geometry are axioms, points, angles, planes and curves. There are several divisions within geometry such as Euclidean geometry, non-Euclidean geometry, differential geometry, topology, and algebraic geometry. Geometry has applications in other branches of mathematics as well as the fields of physics, art and architecture. It relates geometric curves to algebraic equations, thereby applying algebraic methods to geometric questions. The topics covered in this extensive book deal with the core aspects of geometry. It is a vital tool for all researching and studying this field. This book will serve as a valuable source of reference for graduate and postgraduate students.

A detailed account of the significant topics covered in this book is provided below:

Chapter 1- Geometry is a branch of mathematics, which studies the properties, measurements, and relationships of geometric objects. It includes the study of triangles, circles, angles, quadrilaterals, etc. This is an introductory chapter which will briefly introduce about geometry.

Chapter 2- Euclidean geometry is the study and analysis of solid shapes, figures and planes, on the basis of Euclid's axioms, postulates and theorems. Euclidean plane geometry, Euclid's postulates, parallelogram law, Apollonius theorem, etc. fall under its domain. This chapter delves into detailed study of the Euclidean geometry to provide in-depth understanding of it.

Chapter 3- Non-Euclidean geometry is based on principles that are closely related to the Euclidean geometry. It includes elliptic geometry, spherical geometry, hyperbolic geometry. Concepts of hypercycles and horocycles, Apeirogon, tessellations, Gaussian and mean curvatures, etc. are also studied within it. This chapter has been carefully written to provide an easy understanding of non-Euclidean geometry.

Chapter 4- Discrete geometry is the study of geometrical objects and properties that are discrete in their representation. Hyperplanes, polyhedrons, polytope, circle packing, Helly's theorem, etc. come within it. The topics elaborated in this chapter will help in gaining a better perspective about discrete geometry.

Chapter 5- Algebraic geometry is a branch of mathematics which uses algebraic techniques for solving geometrical problems for set of zeros. Some of the concepts related to it are affine variety, Pascal's theorem, Reimann-Roch theorem, etc. This chapter closely examines about these concepts of algebraic geometry to provide an extensive understanding of the subject.

Chapter 6- Analytic geometry is a branch of algebra that deals with the modeling of geometric objects such as points, lines and circles by using a Cartesian system. Conic sections and tangents and normal are also studied under it. This chapter sheds light on analytic geometry for a thorough understanding of the subject.

It gives me an immense pleasure to thank our entire team for their efforts. Finally in the end, I would like to thank my family and colleagues who have been a great source of inspiration and support.

April Watson

1

An Introduction to Geometry

Geometry is a branch of mathematics, which studies the properties, measurements, and relationships of geometric objects. It includes the study of triangles, circles, angles, quadrilaterals, etc. This is an introductory chapter which will briefly introduce about geometry.

Geometry is the branch of mathematics concerned with the shape of individual objects, spatial relationships among various objects and the properties of surrounding space.

Major Branches of Geometry

Euclidean Geometry

In several ancient cultures there developed a form of geometry suited to the relationships between lengths, areas and volumes of physical objects. This geometry was codified in Euclid's *Elements* about 300 BCE on the basis of 10 axioms, or postulates, from which several hundred theorems were proved by deductive logic. The *Elements* epitomized the axiomatic-deductive method for many centuries.

Analytic Geometry

Analytic geometry was initiated by the French mathematician René Descartes, who introduced rectangular coordinates to locate points and to enable lines and curves to be represented with algebraic equations. Algebraic geometry is a modern extension of the subject to multidimensional and non-Euclidean spaces.

Projective Geometry

Projective geometry originated with the French mathematician Girard Desargues to deal with those properties of geometric figures that are not altered by projecting their image, or "shadow," onto another surface.

Differential Geometry

The German mathematician Carl Friedrich Gauss, in connection with practical problems of surveying and geodesy, initiated the field of differential geometry. Using differential calculus, he characterized the intrinsic properties of curves and surfaces. For instance, he showed that the intrinsic curvature of a cylinder is the same as that of a plane, as can be seen by cutting a cylinder along its axis and flattening, but not the same as that of a sphere, which cannot be flattened without distortion.

Non-euclidean Geometries

Beginning in the 19th century, various mathematicians substituted alternatives to Euclid's parallel postulate, which, in its modern form, reads, "given a line and a point not on the line, it is possible to draw exactly one line through the given point parallel to the line." They hoped to show that the alternatives were logically impossible. Instead, they discovered that consistent non-Euclidean geometries exist.

Topology

Topology, the youngest and most sophisticated branch of geometry, focuses on the properties of geometric objects that remain unchanged upon continuous deformation shrinking, stretching, and folding, but not tearing. The continuous development of topology dates from 1911, when the Dutch mathematician L.E.J. Brouwer introduced methods generally applicable to the topic.

Point

A point is the fundamental element of geometry. If we put the tip of a pencil on a paper and press it lightly, we see a very small mark on the paper. This mark is known as a point.

The smallest mark denotes the existence of a point. We say that a point does not have any length, breadth or thickness, but it has a definite position.

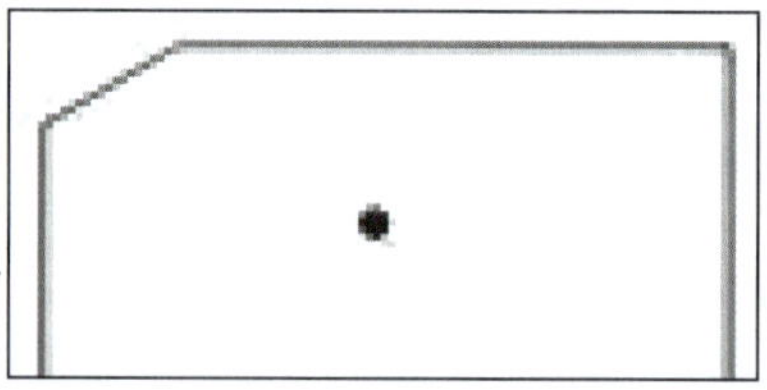

It can be named by a capital letter of the English alphabet like A, B, C, ... etc.

Line and its Types

A Line is a one-dimensional geometrical figure having negligible breadth extending infinitely in both directions. A line may be straight or curved. A straight line has same direction throughout its whole length from point to point. A curved line changes it's direction continuously from point to point.

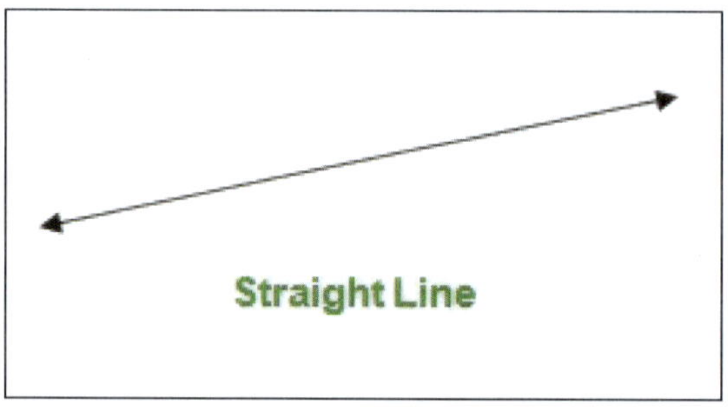

Line Segment

A line segment is a part of a line having specified end points. There can be infinite number of line segments on a line. In the following diagram, AB is a denoted line segment. Two line segments having equal length are called congruent line segments.

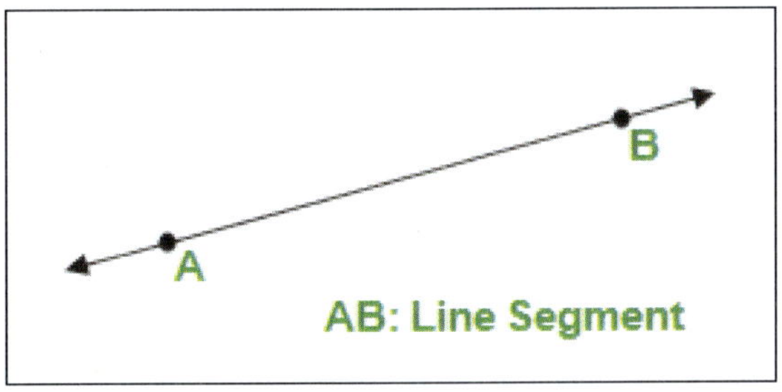

Types of Lines

Horizontal Lines are parallel to the x-axis of the two dimensional plane.

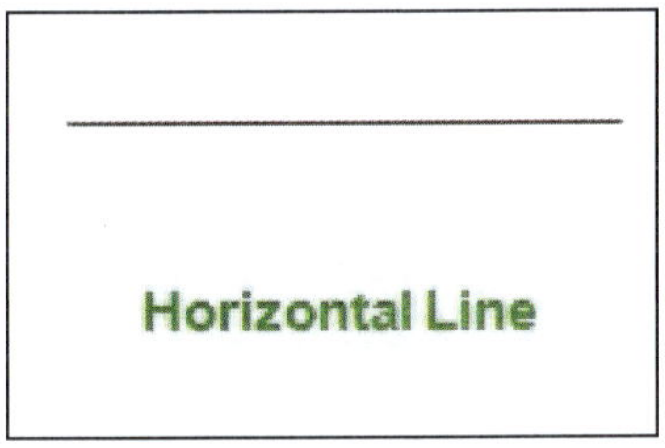

Vertical Lines are parallel to the y-axis of the two-dimensional plane.

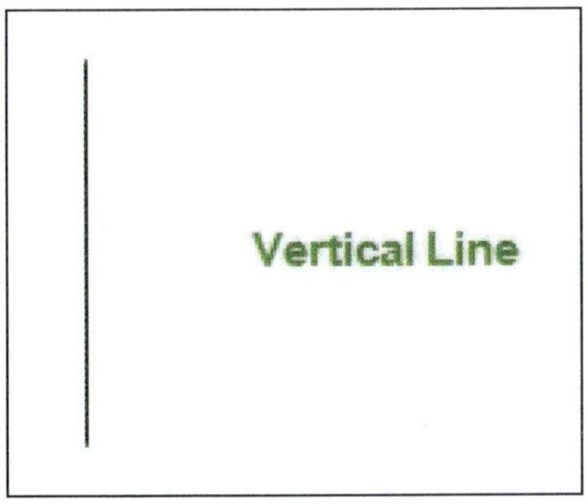

Parallel Lines – If the distance between two straight lines remain same throughout their length at all points, they are said to be parallel. In other words, they do not intersect anywhere in the plane.

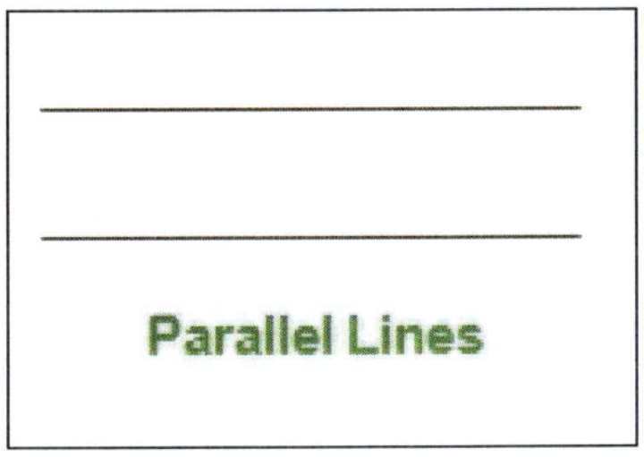

Perpendicular Lines – If a horizontal and vertical line intersect each other on a plane, they are perpendicular.

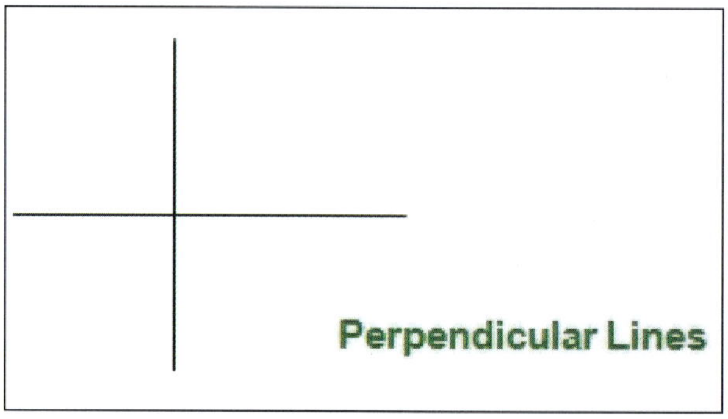

Angle

Angle is the combination of two rays with a common endpoint. The symbol for an angle is ∠. The corner point of an angle is called the vertex of an angle. The two straight sides of the angle are called arms of the angle.

The angles which are measured anticlockwise from the base are something we call as positive angles. The angles measured clockwise from the base are called as negative angles. The standard unit of measuring an angle is degree. It is denoted by °(degree).

Types of Angles

Right Angle

An angle whose measure is equal to 90° is called a right angle. A right angle is formed when two sides make 900. A full rotation consists of four right angles.

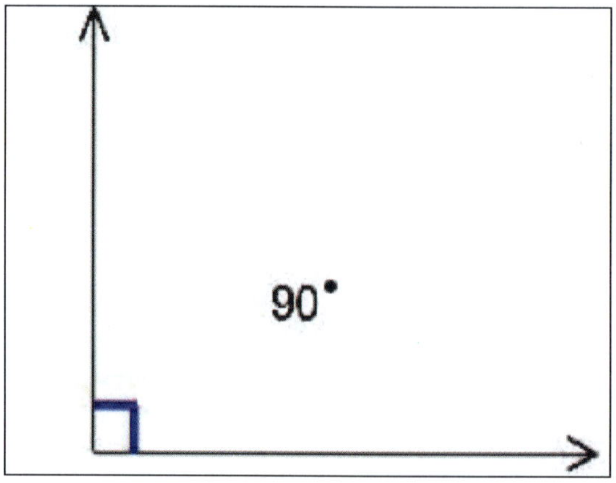

Acute Angle

An angle that is less than 90 degrees is called an acute angle.

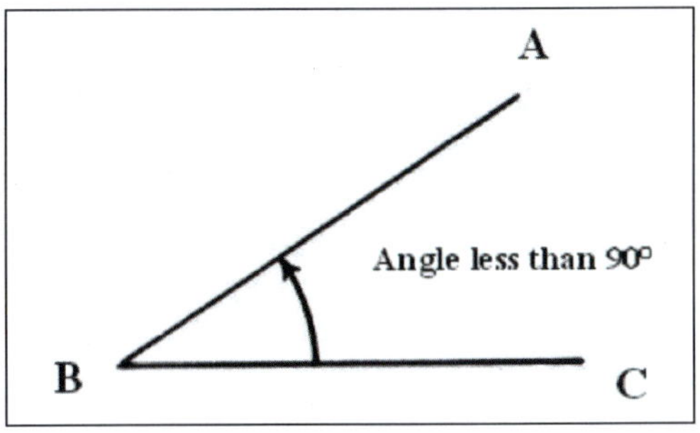

Obtuse Angle

An angle that is between 90 and 180 degrees is called an obtuse angle.

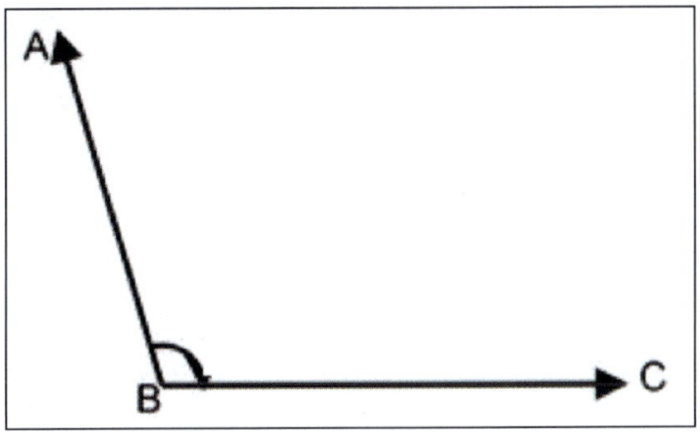

Reflex Angle

An angle that is greater than 180 degrees is called a reflex angle.

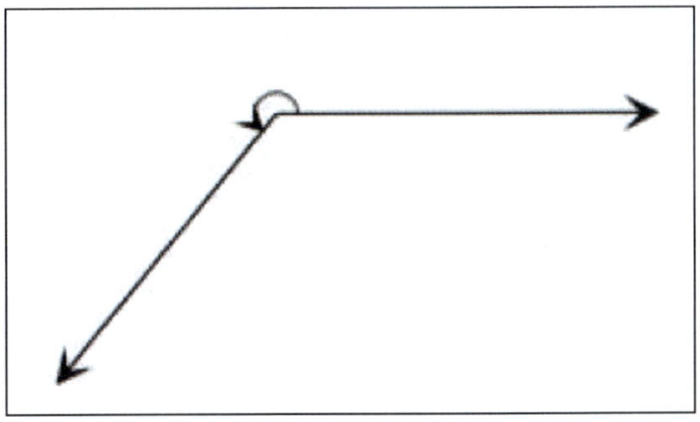

Straight Angle

A straight angle is an internal angle which is equal to 180°.

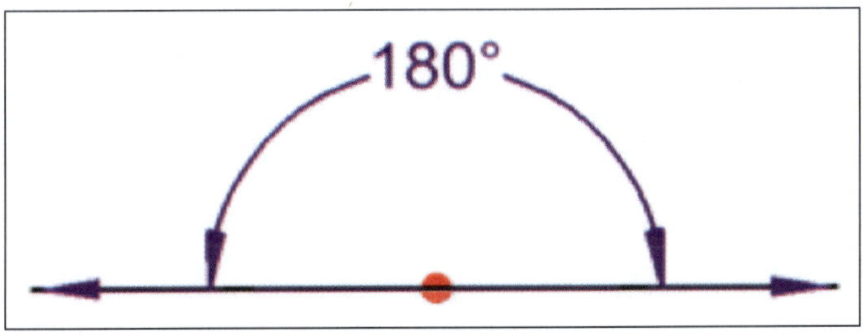

Complete Angle

The angles in a complete turn add up to 360 degrees.

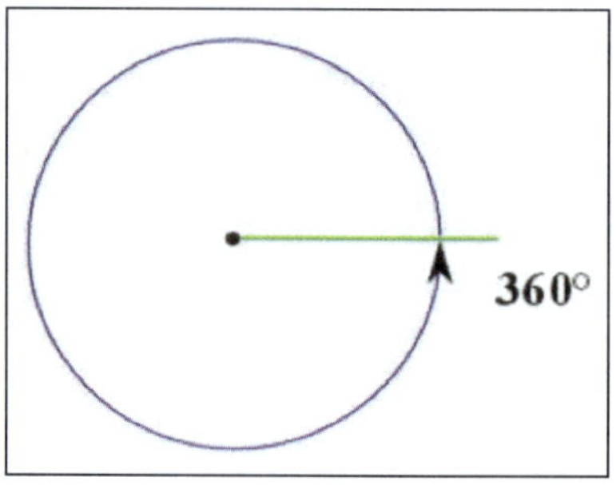

Paired Angles

Adjacent Angles

Two angles that share the same vertex and one side are called adjacent angle. Adjacent angles are two angles that are side by side and share a common ray. They have a common vertex. Any angle can have adjacent angles on its both sides.

Vertically Opposite Angles

Two angles that have a common vertex and opposite to each other formed by the same two lines are called vertically opposite angles. The two vertically opposite angles are always equal. In the diagram below the two angles α (alpha), and β (beta) are vertical angles and $\alpha = \beta$.

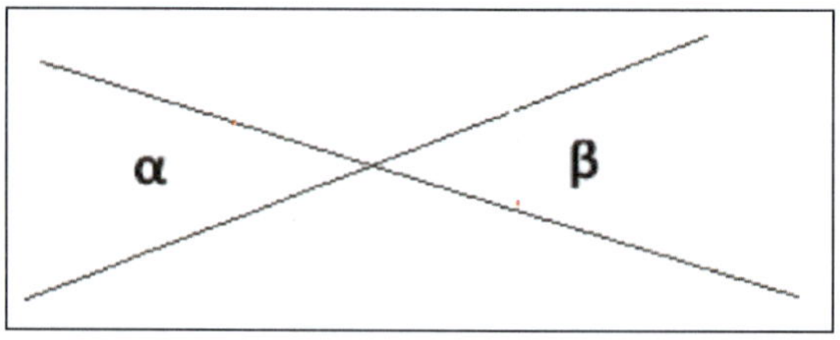

Complementary and Supplementary Angles

Complementary Angles

Two angles whose sum add to 90 degrees are called Complementary Angles.

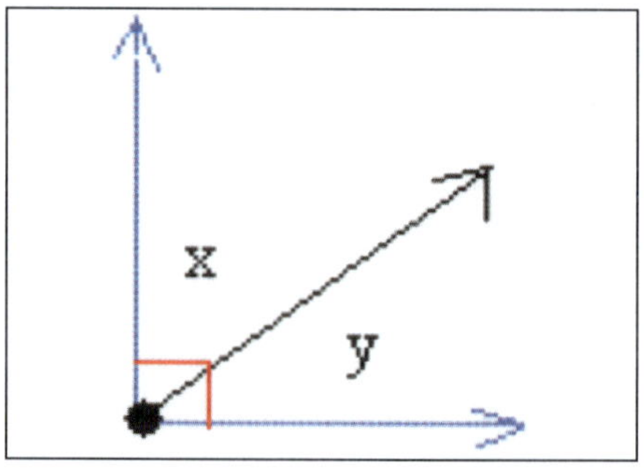

The two complementary angles not necessarily need to be adjacent angles. The following angles are also complementary as the sum of their measures is equal 90 degrees:

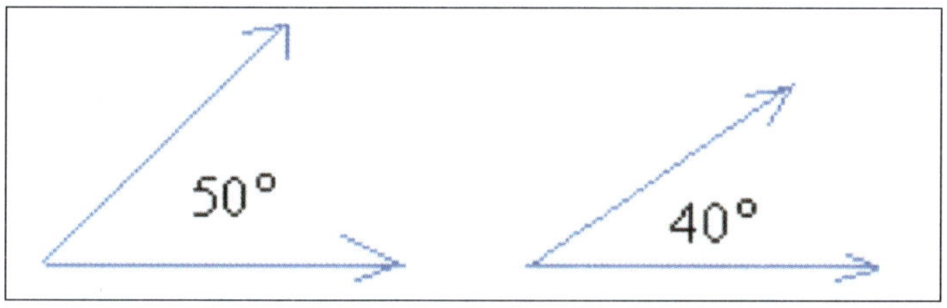

Supplementary Angles

Two angles whose measures add to 180 degrees are called Supplementary Angles. In the figure below $\alpha + \beta = 180$ degree, angles α and β are Supplementary Angles.

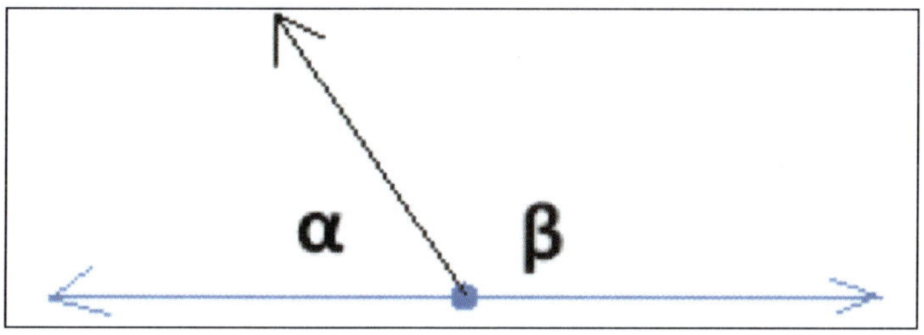

The two supplementary angles need not necessarily be adjacent angles. The following angles are also supplementary as the sum of their measures is equal 180 degrees.

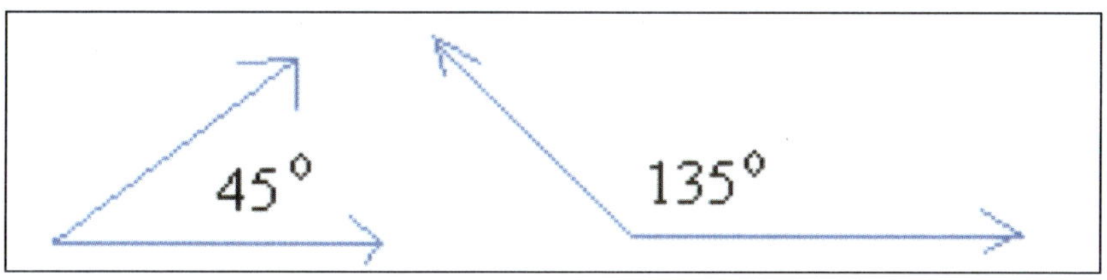

Interior Angle and Exterior Angle

In the figure below, ABC is a geometrical shape called Triangle (we will know more about triangle in other article). Line BC has been extended as line CD. An Interior Angle is an angle inside the shape ABC. The Exterior Angle is the angle between a side of a shape AC, and the extended line CD.

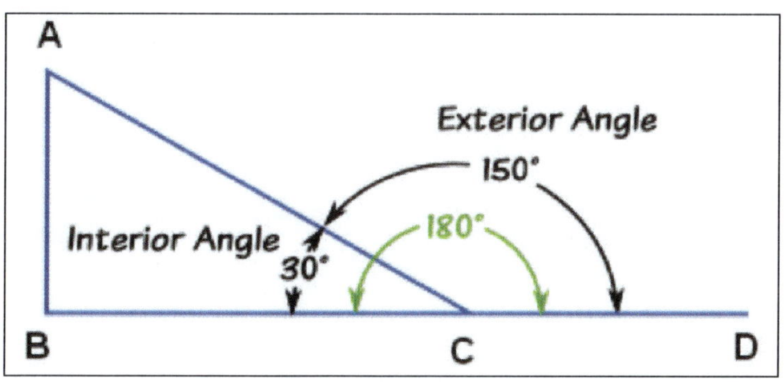

The sum of the Interior Angle and the Exterior Angle are 180°. The two angles are supplementary angles. Here they are adjacent angles also.

Triangles

A triangle is a simple closed curve or polygon which is created by three line-segments. In Euclidean geometry, any three points, specifically non-collinear, form a unique triangle and separately, a unique plane (known as two-dimensional Euclidean space).

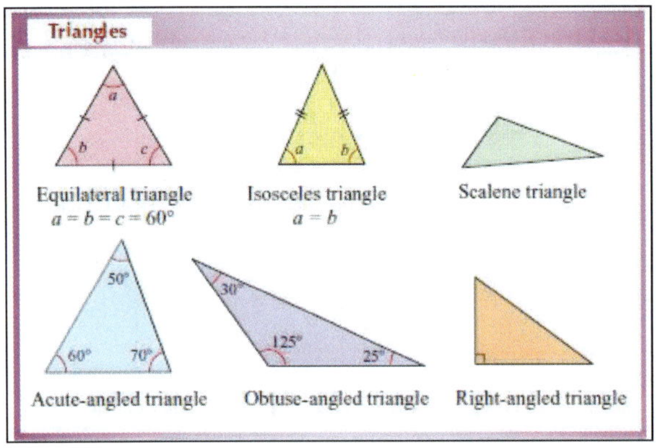

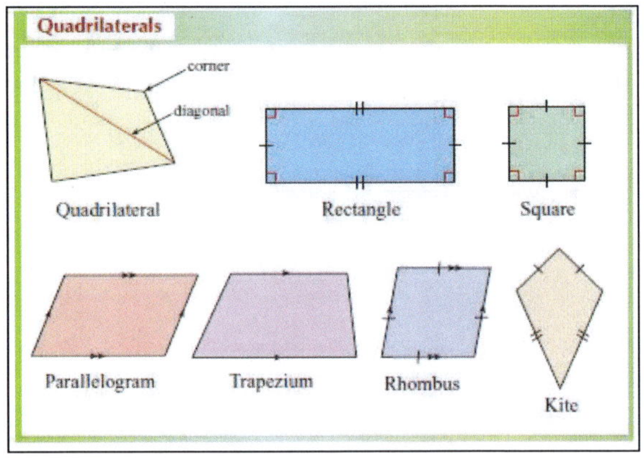

On the other hand, in terms of Euclidean plane geometry, a polygon having four edges (or sides) together with four vertices is called a quadrilateral. At times, the term quadrangle can be used and sometimes tetragon for uniformity with pentagon (5-sided) or hexagon (6-sided).

Types of Triangles

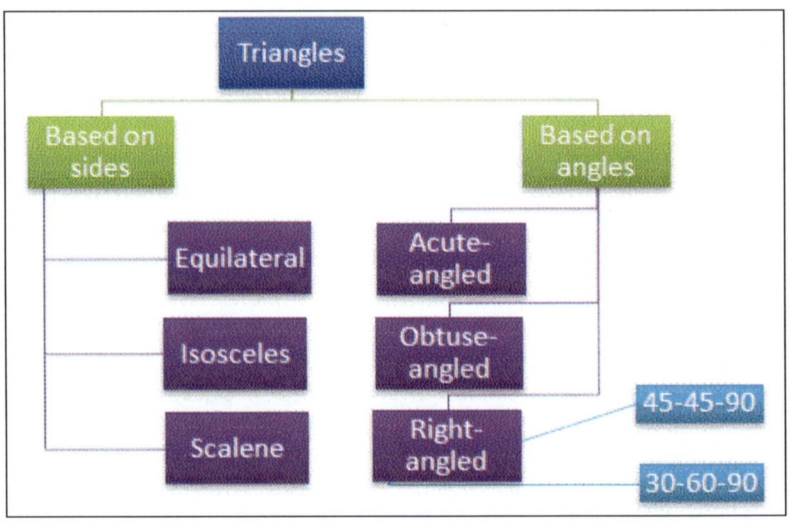

Types of Triangles based on Sides

Equilateral triangle: A triangle having all the three sides of equal length is an equilateral triangle.

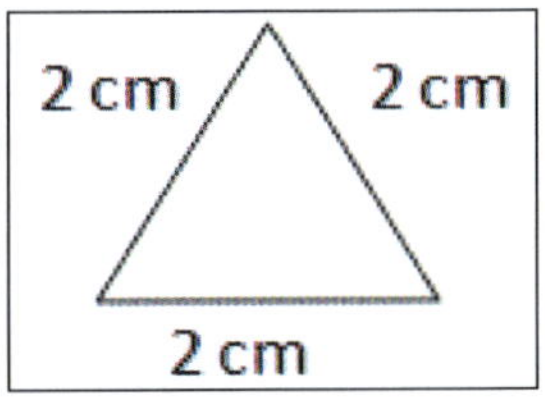

Since all sides are equal, all angles are equal too.

Isosceles triangle: A triangle having two sides of equal length is an Isosceles triangle.

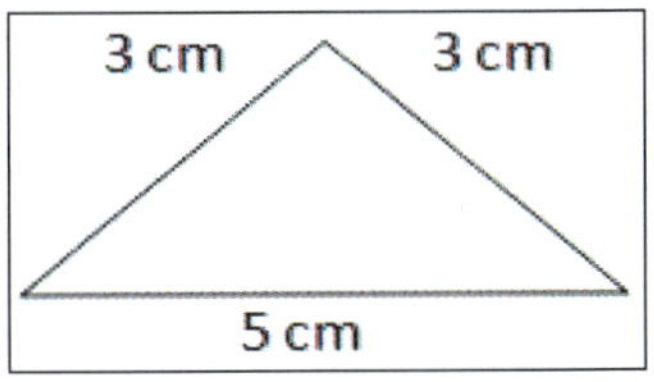

The two angles opposite to the equal sides are equal.

Scalene triangle: A triangle having three sides of different lengths is called a scalene triangle.

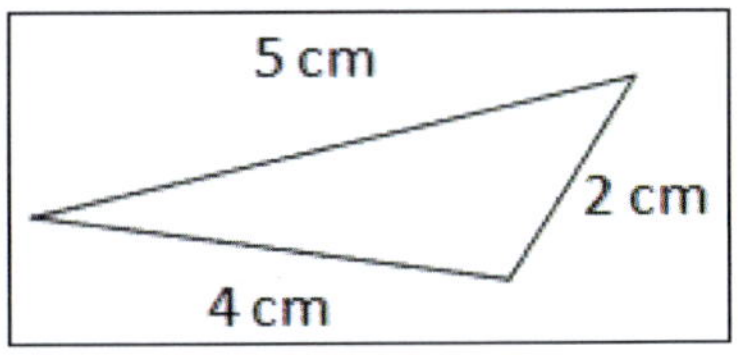

Types of Triangles based on Angles

1. Acute-angled triangle: A triangle whose all angles are acute is called an acute-angled triangle or Acute triangle.

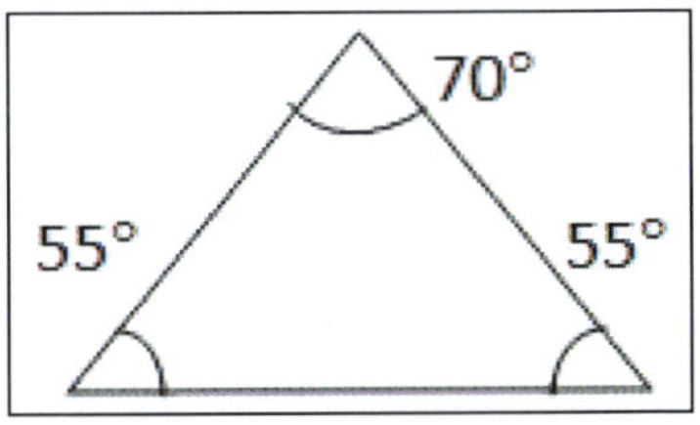

2. Obtuse-angled triangle: A triangle whose one angle is obtuse is an obtuse-angled triangle or Obtuse triangle.

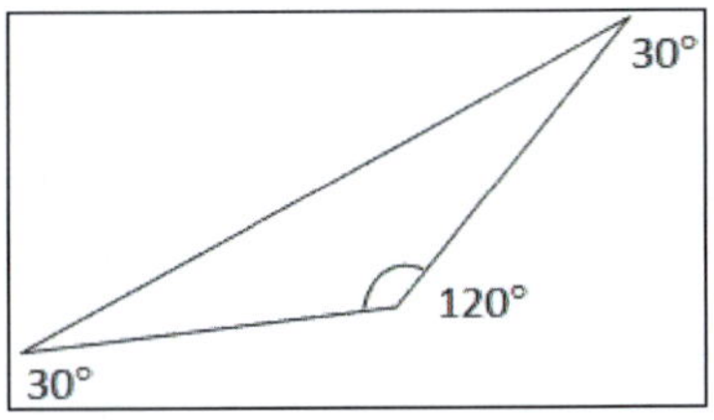

3. Right-angled triangle: A triangle whose one angle is a right-angle is a Right-angled triangle or Right triangle.

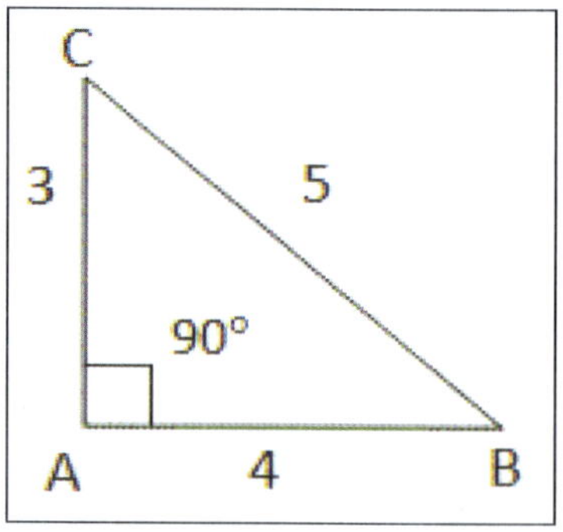

In the figure above, the side opposite to the right angle, BC is called the hypotenuse.

For a Right triangle ABC,

$$BC^2 = AB^2 + AC^2$$

This is called the Pythagorean Theorem. In the triangle above, $5^2 = 4^2 + 3^2$. Only a triangle that satisfies this condition is a right triangle. Hence, the Pythagorean Theorem helps to find whether a triangle is Right-angled.

Types of Right Triangles

There are different types of right triangles. As of now, our focus is only on a special pair of right triangles:

1. 45-45-90 triangle,
2. 30-60-90 triangle.

45-45-90 Triangle

A 45-45-90 triangle, as the name indicates, is a right triangle in which the other two angles are 45° each.

This is an isosceles right triangle.

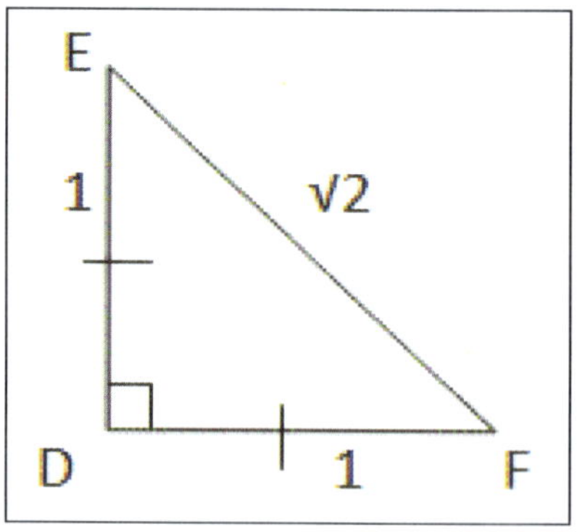

In Δ DEF, DE = DF and $\angle D = 90°$.

The sides in a 45-45-90 triangle are in the ratio $1 : 1 : \sqrt{2}$.

30-60-90 Triangle

A 30-60-90 triangle, as the name indicates, is a right triangle in which the other two angles are 30° and 60°.

This is a scalene right triangle as none of the sides or angles are equal.

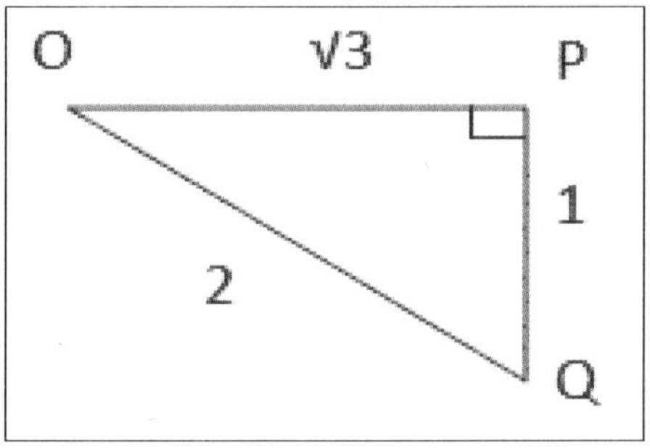

The sides in a 30-60-90 triangle are in the ratio 1 : √3 : 2.

Basic Properties of Triangles

- The sum of the angles in a triangle is 180°. This is called the angle-sum property.
- The sum of the lengths of any two sides of a triangle is greater than the length of the third side. Similarly, the difference between the lengths of any two sides of a triangle is less than the length of the third side.
- The side opposite to the largest angle is the longest side of the triangle and the side opposite to the smallest angle is the shortest side of the triangle.

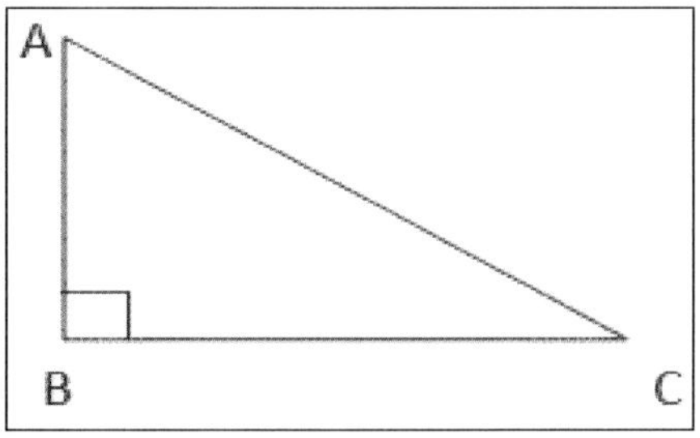

In the figure above, ∠B is the largest angle and the side opposite to it (hypotenuse), is the largest side of the triangle.

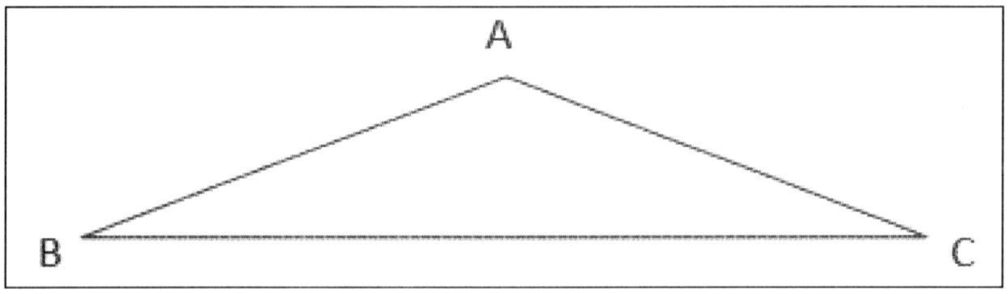

In the figure above, ∠A is the largest angle and the side opposite to it, BC is the largest side of the triangle.

- An exterior angle of a triangle is equal to the sum of its interior opposite angles. This is called the exterior angle property of a triangle.

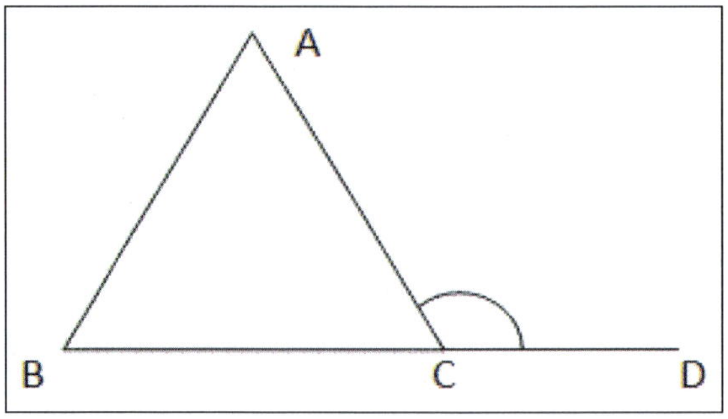

Here, $\angle ACD$ is the exterior angle to the ΔABC.

According to the exterior angle property, $\angle ACD = \angle CAB + \angle ABC$.

Triangle Inequalities

The triangle inequality states that the sum of the lengths of any two sides of a triangle is greater than the length of the remaining side.

It follows from the fact that a straight line is the shortest path between two points. The inequality is strict if the triangle is non-degenerate (meaning it has a non-zero area).

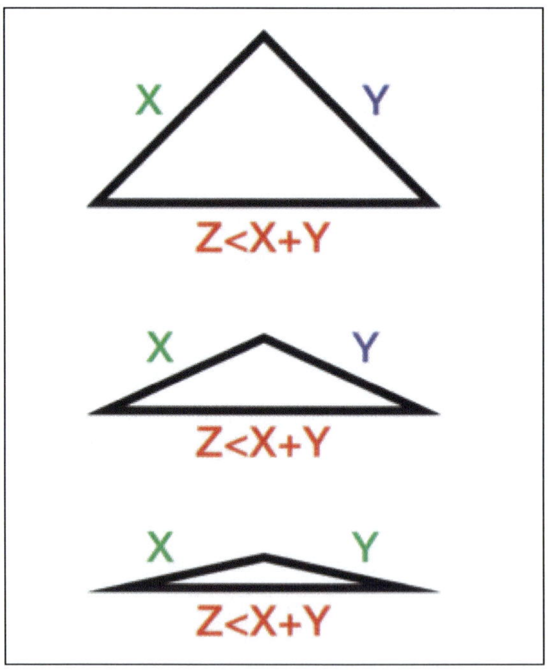

If two side lengths of a non-degenerate triangle are each equal to 66, what are the possible values for the length of the third side?

Let the third side have length x, then from the triangle inequality we have $6+6>x$ and $6+x>6$, implying $0<x<12$. So any real number x such that $0<x<12$ is a possible length of the third side.

If two side lengths of a non-degenerate triangle are 9 and 15, what are the possible lengths of the third side?

Let the third side have length $x>0$. The 3 sides must satisfy:

$x<9+15$,

$15<x+9$,

$9<x+15$.

The third inequality is trivially true, and the first two give $6<x<24$. square.

If the three integer sides of a triangle are given by $x+2$, $2x+7$, $4x+1$, what is the greatest possible value of x?

By the triangle inequality we have:

$(x+2)+(2x+7)>(4x+1) \Rightarrow x<8$

$(x+2)+(4x+1)>(2x+7) \Rightarrow x> 4/3$

$(2x+7)+(4x+1)>(x+2)\ x> -6/5$

which implies $4/3< x<8$. So the greatest possible integer value of x is 7.

Given 55 sticks of respective lengths 1,3,5,9 and 10, how many distinct triangles can be formed from three of the sticks?

If one side of the triangle has a side length of 1, then there is no possible combination to form a triangle. In fact, if the three sides of a triangle have distinct integer lengths, then it is impossible to have one side of unit length.

- If one side has a length of 3, the only possible combination is (3,9,10).
- If one side has a length of 5, the only possible combination is 5,9,10).
- If one side has a length of 9, the possible combinations are (9,3,10) and (9,5,10).
- If one side has a length of 10, the possible combinations are (10,9,3) and (10,9,5).

Since changing the order of the sides doesn't change the triangle itself, we can form 2 distinct triangles.

Vectors

The triangle inequality has the following formulation in terms of vectors in $\mathbb{R}^n$.

Let x and y be vectors in $\mathbb{R}^n$. Let $||.||$ denote the norm of a vector. Then,

$$||x+y|| \le ||x|| + ||y||.$$

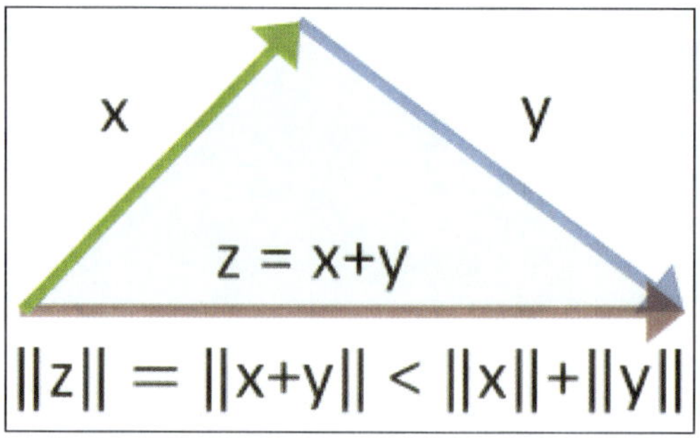

$$(\|x\|+\|y\|)^2-\|x+y\|^2=\|x\|^2+\|y\|^2+2\|x\|\|y\|-(x+y)\cdot(x+y)$$

$$=\|x\|^2+\|y\|^2+2\|x\|\|y\|-(x\cdot x)-(y\cdot y)-2(x\cdot y)$$

$$=2\|x\|\|y\|-2(x\cdot y)$$

$$\geq 0$$

Note that equality holds in Cauchy-Schwarz if and only if the vectors are parallel, which is the condition that the triangle is degenerate.

The above proof is essentially the same as using the law of cosines: if a, b, c are the sides of a triangle:

$$c^2=a^2+b^2-2abcos(\theta)$$

and the right side is maximized when $\cos(\theta)$ is minimized, i.e. when it is -1, so

$$c^2\leq a^2+b^2-2ab(-1)=(a+b)^2.$$

Therefore, $c\leq a+b$ with equality if and only if the angle between the sides of lengths a and b is 180°.

Metric Spaces

The triangle inequality is a fundamental property of generalized distance functions called metrics, which are used to construct metric spaces. A metric is a function $d(x,y)$ which takes two arguments from a set X and produces a nonnegative real number, with the following properties:

$$d(x,y)=0 \text{ if and only if } x=y.$$
$$d(x,y)=d(y,x).$$
$$d(x,y)+d(y,z)\geq d(x,z).$$

This third property is the triangle inequality, and it is important in proving many of the topological properties of metric spaces.

Similarity of Triangles

Two triangles are said to be similar when they have two corresponding angles congruent and the sides proportional.

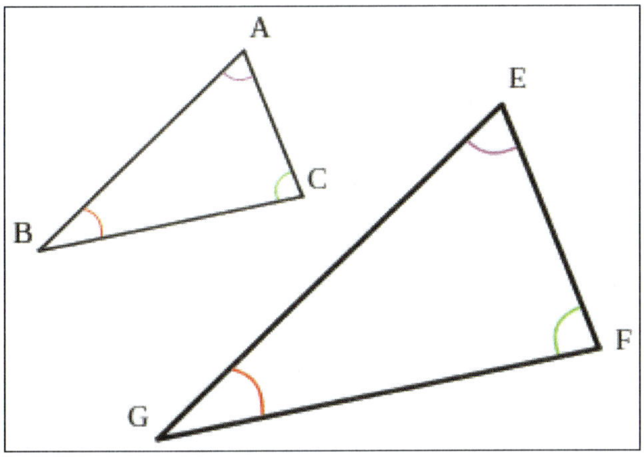

In the above diagram, we see that triangle EFG is an enlarged version of triangle ABC i.e., they have the same shape. The same shape of the triangle depends on the angle of the triangles.

$\angle ABC = \angle EGF$, $\angle BAC = \angle GEF$, $\angle EFG = \angle ACB$.

The area, altitude, and volume of Similar triangles are in the same ratio as the ratio of the length of their sides.

Properties of similar triangles:

- Reflexivity: A triangle (Δ) is similar to itself.
- Symmetry: If Δ ABC ~ Δ DEF, Then Δ DEF ~ Δ ABC.
- Transitivity: IfΔ ABC ~Δ DEF andΔ DEF ~Δ XYZ, then Δ ABC ~Δ XYZ.

Tests to Prove that a Triangle is Similar

Angle-angle Similarity (AA)

If two corresponding angles of the two triangles are congruent, the triangle must be similar.

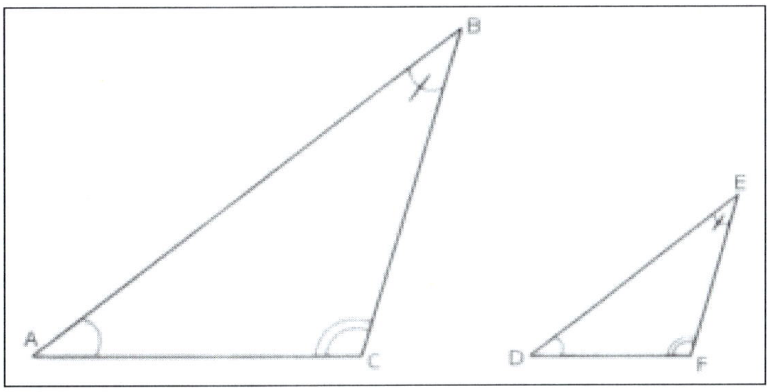

Side-side-side Similarity (SSS)

If the corresponding sides of the two triangles are proportional the triangles must be similar.

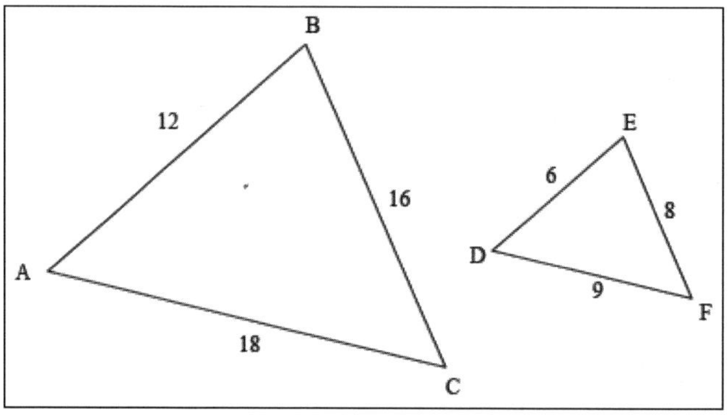

Side Angle Side Similarity (SAS)

If two sides of two triangles are proportional and they have one corresponding angle congruent, the two triangles are said to be similar.

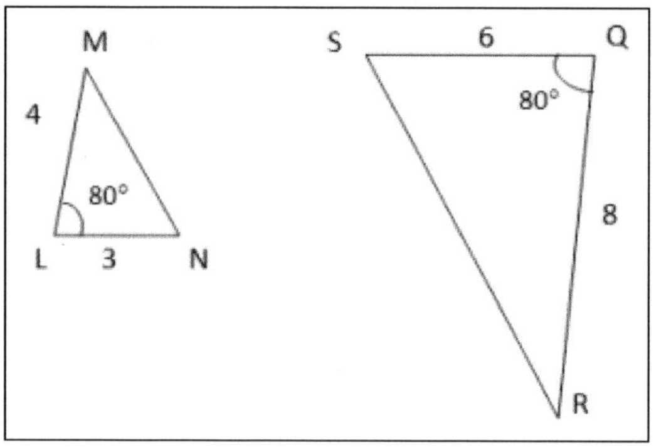

Theorem

The ratio of the areas of two similar triangles is equal to the square of the ratio of their corresponding sides.

Given: A(ΔABC)~A(ΔPQR).

To Prove: A(ΔABC)/A(ΔPQR)=AB^2/PQ^2.

Construction: Construct seg AM perpendicular side BC and seg PN perpendicular side QR

Proof:

A(ΔABC)/A(ΔPQR)=(BC)(AM)/(QR)(PN)

A(ΔABC)~A(ΔPQR).

Therefore, ∠B=∠Q

AB/PQ=BC/QR=AC/PR.

In triangle ABM and triangle PQN,

∠ABM=∠PQN.

∠AMB=∠PNQ (each side is a right angle).

Therefore, Δ ABM ~Δ PQN (AA test of similarity):

AB/PQ=AM/PN (c.s.s.t)

A(ΔABC)/A(ΔPQR)=[(BC)/(QR)][(AB)/(PQ)]

A(ΔABC)/A(ΔPQR)=[(AB)/(PQ)][(AB)/(PQ)]

A(ΔABC)/A(ΔPQR)=AB^2/PQ^2.

Similarly, we can show that,

A(ΔABC)/A(ΔPQR)=$BC^2/QR^2=AC^2/PR^2$.

Hence, we have proved that the ratio of the areas of two similar triangles is equal to the square of the ratio of their corresponding sides.

Congruence of Triangles

A polygon made of three line segments forming three angles is known as Triangle. Two triangles are said to be congruent if their sides have the same length and angles have same measure. Thus two triangles can be superimposed side to side and angle to angle.

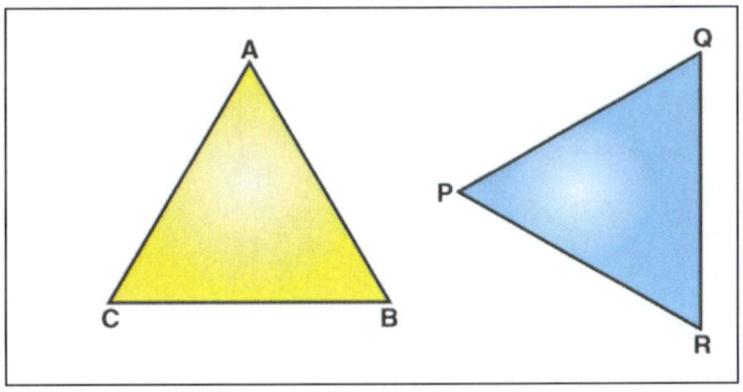

In the above figure, D ABC and D PQR are congruent triangles. This means,

- Vertices: A and P, B and Q, and C and R are same.
- Sides: AB=PQ, QR= BC and AC=PR.
- Angles: Đ A = Đ P, ĐB = ĐQ and Đ C= Đ R.

Congruent triangles are triangles having corresponding sides and angles to be equal. Congruence is denoted by the symbol ≅. They have the same area and the same perimeter.

CPCT in Maths

The full form of CPCT is Corresponding parts of Congruent triangles. Congruency can be predicted without actually measuring the sides and angles of a triangle. Different rules of congruency are as follows.

SSS (Side-side-side)

If all the three sides of one triangle are equivalent to the corresponding three sides of the second triangle, then the two triangles are said to be congruent by SSS rule.

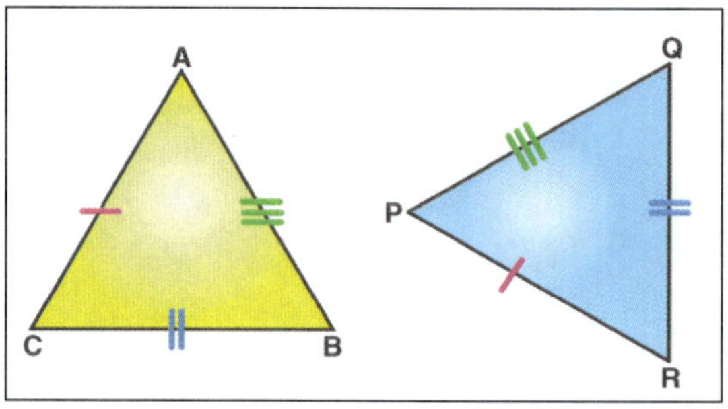

In the above-given figure, AB= PQ, QR= BC and AC=PR,

Hence, D ABC ≅ D PQR.

SAS (Side-angle-side)

If any two sides and angle included between the sides of one triangle are equivalent to the corresponding two sides and the angle between the sides of the second triangle, then the two triangles are said to be congruent by SAS rule.

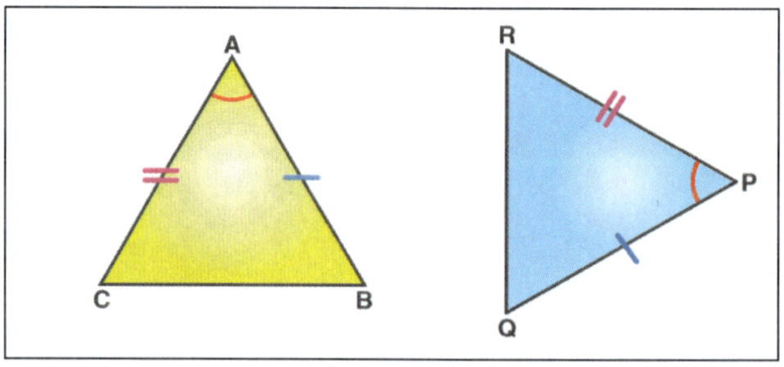

In above given figure, sides AB= PQ, AC=PR and angle between AC and AB equal to angle between PR and PQ i.e. Ð A = Ð P, hence D ABC ≅ D PQR.

ASA (Angle-side-angle)

If any two angles and side included between the angles of one triangle are equivalent to the corresponding two angles and side included between the angles of the second triangle, then the two triangles are said to be congruent by ASA rule.

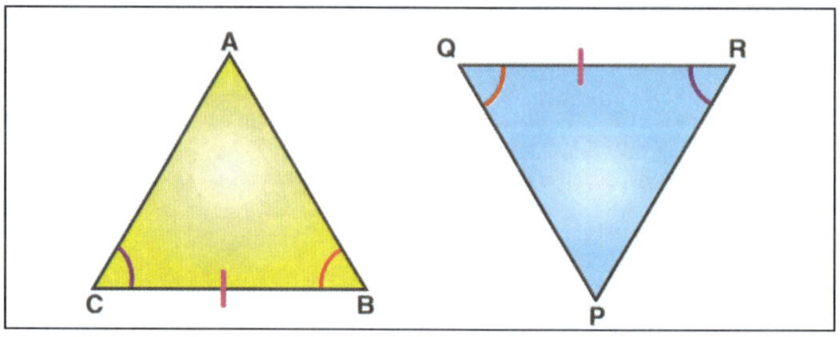

In above given figure, Đ B = Đ Q, Đ C = Đ R and sides between Đ B and ĐC, Đ Q and Đ R are equal to each other i.e. BC= QR, hence D ABC ≅ D PQR.

RHS (Right Angle-hypotenuse-side)

If the hypotenuse and a side of a right- angled triangle is equivalent to the hypotenuse and a side of the second right- angled triangle, the right triangles are said to be congruent by RHS rule.

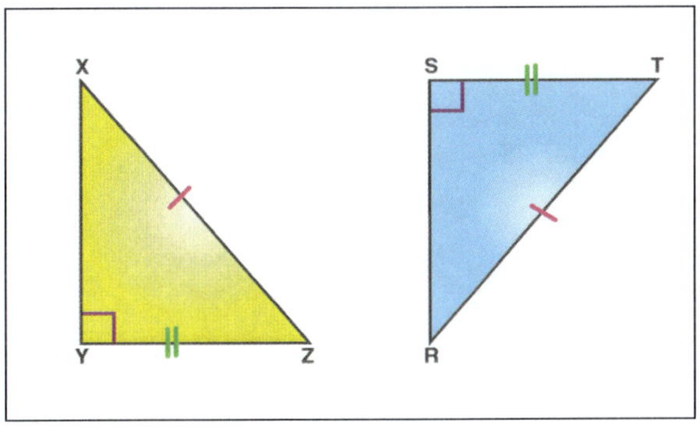

In above figure, hypotenuse XZ = RT and side YZ=ST, hence triangle XYZ ≅ triangle RST.

Example: In the following figure, AB = BC and AD = CD. Show that BD bisects AC at right angles.

Solution: We are required to prove $\angle BEA = \angle BEC = 90°$ and AE = EC.

Consider ΔABD and ΔCBD,

- AB = BC (Given)
- AD = CD (Given)
- BD = BD (Common)

Therefore, ΔABD ≅ ΔCBD (By SSS congruency),

- $\angle ABD = \angle CBD$ (CPCTC)

Now, consider ΔABE and ΔCBE,

- AB = BC (Given)
- $\angle ABD = \angle CBD$ (Proved above)
- BE = BE (Common)

Therefore, ΔABE≅ ΔCBE (By SAS congruency),

- $\angle BEA = \angle BEC$ (CPCTC)
- And $\angle BEA + \angle BEC = 180°$ (Linear pair)
- $2\angle BEA = 180°$ ($\angle BEA = \angle BEC$)

- $\angle BEA = 180^\circ/2 = 90^\circ = \angle BEC$
- AE = EC (CPCTC)

Hence, BD is perpendicular bisector of AC.

Circle

In Euclidean geometry, a circle is the set of all points in a plane at a fixed distance, called the radius, from a given point, the center. The length of the circle is called its circumference, and any continuous portion of the circle is called an arc.

A circle is a simple closed curve that divides the plane into an interior and exterior. The interior of the circle is called a disk.

Mathematically, a circle can be understood in several other ways as well. For instance, it is a special case of an ellipse in which the two foci coincide (that is, they are the same point). Alternatively, a circle can be thought of as the conic section attained when a right circular cone is intersected with a plane perpendicular to the axis of the cone.

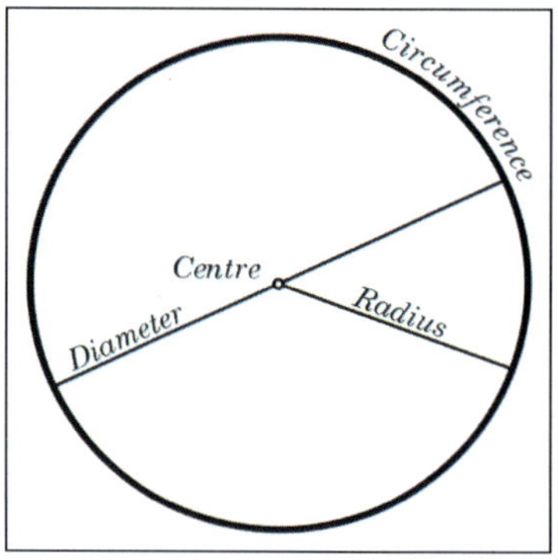

Properties

All circles have similar properties. Some of these are noted below:

- For any circle, the area enclosed and the square of its radius are in a fixed proportion, equal to the mathematical constant π.
- For any circle, the circumference and radius are in a fixed proportion, equal to 2π.
- The circle is the shape with the highest area for a given length of perimeter.
- The circle is a highly symmetrical shape. Every line through the center forms a line of reflection symmetry. In addition, there is rotational symmetry around the center for every angle. The symmetry group is called the orthogonal group O(2,R), and the group of rotations alone is called the circle group T.
- The circle centered at the origin with radius 1 is called the unit circle.

Chord Properties

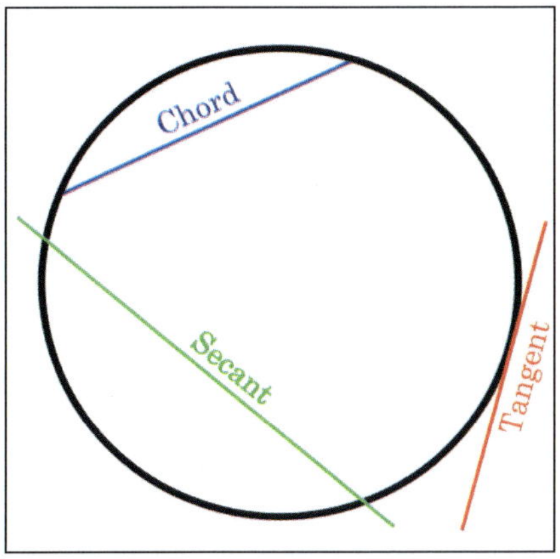

A line segment that connects one point of a circle to another is called a *chord*. The *diameter* is a chord that runs through the center of the circle.

- The diameter is longest chord of the circle.
- Chords equidistant from the center of a circle are equal in length. Conversely, chords that are equal in length are equidistant from the center.
- A line drawn through the center of a circle perpendicular to a chord bisects the chord. Alternatively, one can state that a line drawn through the center of a circle bisecting a chord is perpendicular to the chord. This line is called the *perpendicular bisector* of the chord. Thus, one could also state that the perpendicular bisector of a chord passes through the center of the circle.
- If a central angle and an inscribed angle of a circle are subtended by the same chord and on the same side of the chord, then the central angle is twice the inscribed angle.
- If two angles are inscribed on the same chord and on the same side of the chord, then they are equal.
- If two angles are inscribed on the same chord and on opposite sides of the chord, then they are supplemental.
- An inscribed angle subtended by a diameter is a right angle.

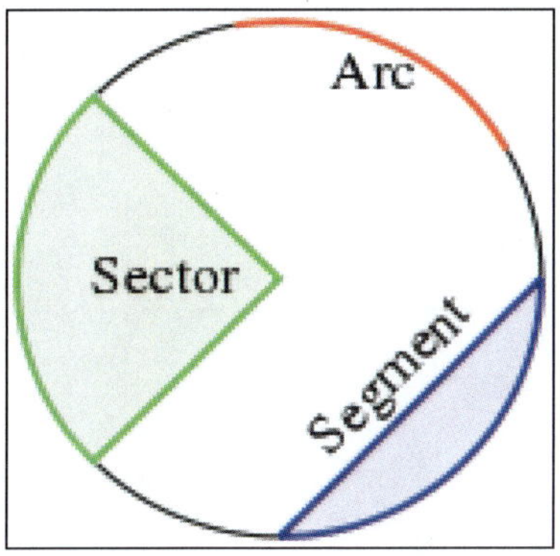

Sagitta Properties

The sagitta is a line segment drawn perpendicular to a chord, between the midpoint of that chord and the circumference of the circle.

Given the length of a chord, y, and the length x of the sagitta, the Pythagorean theorem can be used to calculate the radius of the unique circle which will fit around the 2 lines:

$$r = \frac{y^2}{8x} + \frac{x}{2}$$

Tangent Properties

- The line drawn perpendicular to the end point of a radius is a tangent to the circle.
- A line drawn perpendicular to a tangent at the point of contact with a circle passes through the center of the circle.
- Tangents drawn from a point outside the circle are equal in length.
- Two tangents can always be drawn from a point outside of the circle.

Theorems

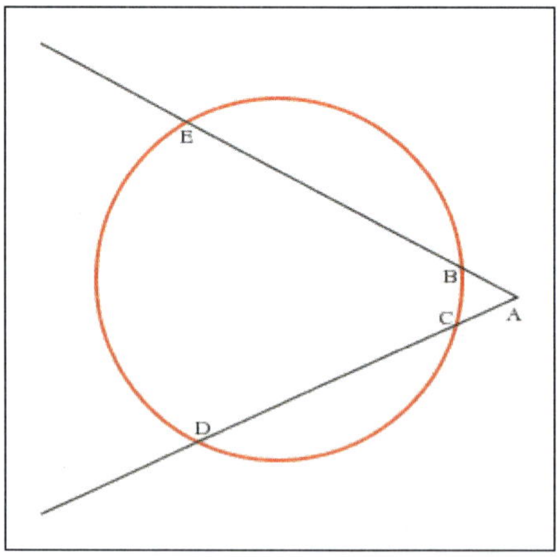

Secant-secant theorem.

- The chord theorem states that if two chords, CD and EF, intersect at G, then $CG \times DG = EG \times FG$ (Chord theorem).
- If a tangent from an external point D meets the circle at C and a secant from the external point D meets the circle at G and E respectively, then $DC^2 = DG \times DE$ (tangent-secant theorem).
- If two secants, DG and DE, also cut the circle at H and F respectively, then $DH \times DG = DF \times DE$. (Corollary of the tangent-secant theorem).
- The angle between a tangent and chord is equal to the subtended angle on the opposite side of the chord. (Tangent chord property).

- If the angle subtended by the chord at the center is 90 degrees then l = √(2) × r, where l is the length of the chord and r is the radius of the circle.
- If two secants are inscribed in the circle as shown at right, then the measurement of angle A is equal to one half the difference of the measurements of the enclosed arcs (DE and BC). This is the secant-secant theorem.

Analytic Results

Equation of a Circle

In an x-y coordinate system, the circle with center (a, b) and radius r is the set of all points (x, y) such that:

$$(x-a)^2+(y-b)^2=r^2.$$

If the circle is centered at the origin (0, 0), then this formula can be simplified to:

$$x^2+y^2=r^2$$

and its tangent will be:

$$xx_1+yy_1=r^2$$

Where, x^1, y^1 are the coordinates of the common point.

When expressed in parametric equations, (x, y) can be written using the trigonometric functions sine and cosine as:

$$x=a+r\cos t,$$
$$y=b+r\sin t$$

Where, t is a parametric variable, understood as the angle the ray to (x, y) makes with the x-axis.

In homogeneous coordinates each conic section with equation of a circle is:

$$ax^2+ay^2+2b_1xz+2b_2yz+cz^2=0.$$

It can be proven that a conic section is a circle if and only if the point I(1,i,0) and J(1,-i,0) lie on the conic section. These points are called the circular points at infinity.

In polar coordinates the equation of a circle is:

$$r^2-2rr_0\cos(\theta-\varphi)+r_0^2=a^2.$$

In the complex plane, a circle with a center at c and radius r has the equation $|z-c|^2=r^2$. *Since* $|z-c|^2=z\overline{z}-\overline{c}z-c\overline{z}+c\overline{c}$ the slightly generalized equation $pz\overline{z}+gz+\overline{gz}=q$ for real p, q and complex g is sometimes called a generalized circle. It is important to note that not all generalized circles are actually circles.

Slope

The slope of a circle at a point (x, y) can be expressed with the following formula, assuming the center is at the origin and (x, y) is on the circle:

$$y' = -\frac{x}{y}.$$

More generally, the slope at a point (x, y) on the circle $(x-a)^2 + (y-b)^2 = r^2$ (i.e., the circle centered at [a, b] with radius r units), is given by:

$$y' = \frac{a-x}{y-b},$$

provided that $y \neq b$, of course.

Area Enclosed

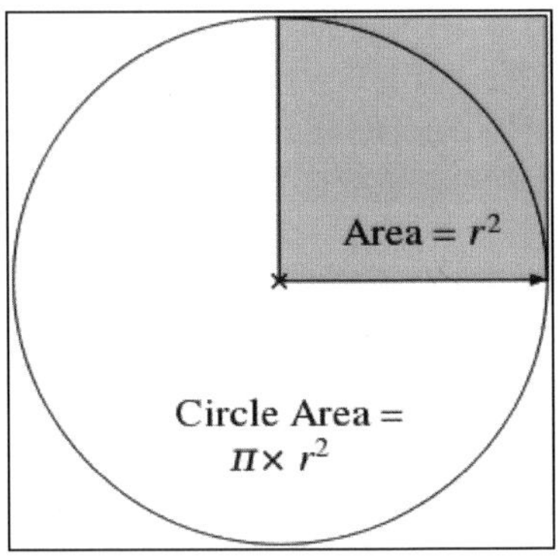

Area of the circle = π × area of the shaded square.

The area enclosed by a circle is:

$$A = r^2 \cdot \pi = \frac{d^2 \cdot \pi}{4} \approx 0.7854 \cdot d^2,$$

that is, approximately 79 percent of the circumscribed square.

Circumference

Length of a circle's circumference is:

$$c = \pi d = 2\pi \cdot r.$$

Alternate formula for circumference:

Given that the ratio circumference c to the Area A is,

$$\frac{c}{A} = \frac{2\pi r}{\pi r^2}.$$

The r and the π can be canceled, leaving:

$$\frac{c}{A}=\frac{2}{r}.$$

Therefore solving for c:

$$c=\frac{2A}{r}$$

So the circumference is equal to 2 times the area, divided by the radius. This can be used to calculate the circumference when a value for π cannot be computed.

Diameter

The diameter of a circle is:

$$d=2r=2\cdot\sqrt{\frac{A}{\pi}}\approx 1.1284\cdot\sqrt{A}.$$

Inscribed Angles

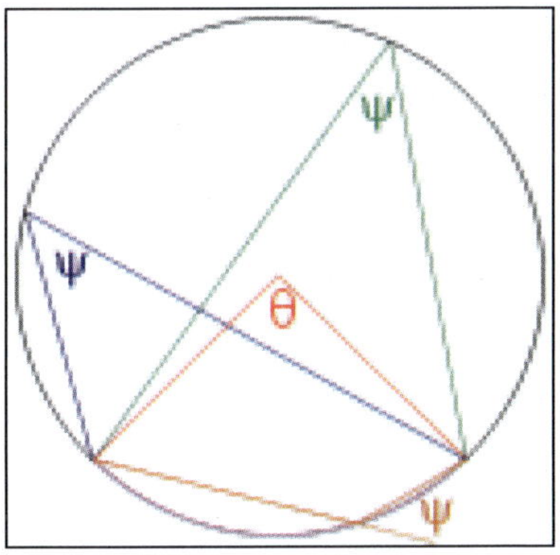

Inscribed angle theorem.

An inscribed angle ψ is exactly half of the corresponding central angle\theta. Hence, all inscribed angles that subtend the same arc have the same value (cf. the blue and green angles \psi in the figure). Angles inscribed on the arc are supplementary. In particular, every inscribed angle that subtends a diameter is a right angle.

An alternative definition of a circle:

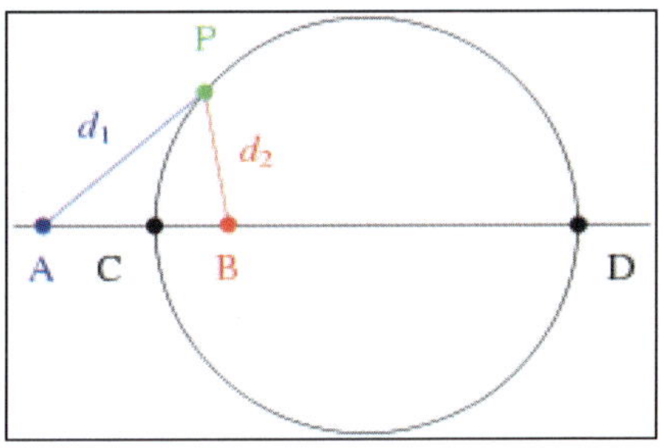

Apollonius of Perga showed that a circle may also be defined as the set of points having a constant ratio of distances to two foci, A and B.

The proof is as follows. A line segment PC bisects the interior angle APB, since the segments are similar:

$$\frac{AP}{BP} = \frac{AC}{BC}$$

Analogously, a line segment PD bisects the corresponding exterior angle. Since the interior and exterior angles sum to 1800 the angle CPD is exactly 900 i.e., a right angle. The set of points P that form a right angle with a given line segment CD form a circle, of which CD is the diameter.

As a point of clarification, note that C and D are determined by A, B, and the desired ratio (i.e. A and B are not arbitrary points lying on an extension of the diameter of an existing circle).

Calculating the Parameters of a Circle

Early science, particularly geometry and astronomy/astrology, was connected to the divine for most medieval scholars. The compass in this thirteenth century manuscript is a symbol of God's act of Creation, as many believed that there was something intrinsically "divine" or "perfect" that could be found in circles.

Given three non-collinear points lying on the circle:

$$\mathrm{P}_1 = \begin{bmatrix} x_1 \\ y_1 \\ z_1 \end{bmatrix}, \mathrm{P}_2 = \begin{bmatrix} x_2 \\ y_2 \\ z_2 \end{bmatrix}, \mathrm{P}_3 = \begin{bmatrix} x_3 \\ y_3 \\ z_3 \end{bmatrix}$$

Radius

The radius of the circle is given by:

$$\mathrm{r} = \frac{|P_1 - P_2||P_2 - P_3||P_3 - P_1|}{2|(P_1 - P_2)\times(P_2 - P_3)|}$$

Center

The center of the circle is given by:

$$\mathrm{P_c} = \alpha P_1 + \beta P_2 + \gamma P_3$$

where,

$$\alpha = \frac{|P_2 - P_3|^2 (P_1 - P_2)\cdot(P_1 - P_3)}{2|(P_1 - P_2)\times(P_2 - P_3)|^2}$$

$$\beta = \frac{|P_1 - P_3|^2 (P_2 - P_1)\cdot(P_2 - P_3)}{2|(P_1 - P_2)\times(P_2 - P_3)|^2}$$

$$\gamma = \frac{|P_1 - P_2|^2 (P_3 - P_1)\cdot(P_3 - P_2)}{2|(P_1 - P_2)\times(P_2 - P_3)|^2}$$

Plane Unit Normal

A unit normal of the plane containing the circle is given by,

$$\hat{n} = \frac{(P_2 - P_1)\times(P_3 - P_1)}{|(P_2 - P_1)\times(P_3 - P_1)|}$$

Parametric Equation

Given the radius, r, center, P_c, a point on the circle, P_0 and a unit normal of the plane containing the circle, n, the parametric equation of the circle starting from the point P_0 and proceeding counterclockwise is given by the following equation:

$$R(s) = P_c + \cos\left(\frac{s}{r}\right)(P_0 - P_c) + \sin\left(\frac{s}{r}\right)\left[\hat{n}\times(P_0 - P_c)\right]$$

Quadrilaterals

Square

A square is a polygon with 4 sides of equal length and 4 right angle corners (90 degree corners). Because it has 4 sides of equal length, a square is a regular quadrilateral. A square is also a rectangle with equal sides and a rhombus with right angles. The area of a square is equal to the length of one side to the power of two (length squared). The perimeter of a square is 4 times the length of one side. A square has a larger area than all other quadrilaterals with the same perimeter. The diagonals of a square bisect each other at 90 degrees and are perpendicular. Opposite sides of a square are parallel. The internal angles of a square add to 360 degrees. A square has 4 lines of reflectional symmetry. Square shapes are often used by humans for design and engineering purposes such as town planning. A cube features six square faces and in many ways is like a square, only in three dimensions rather than two.

Rectangle

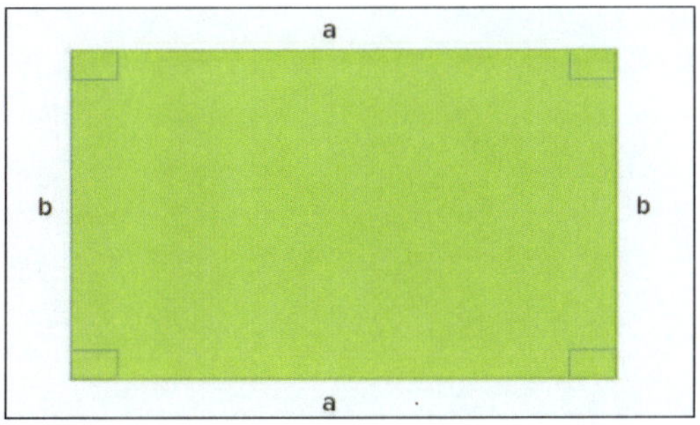

A rectangle is a 2D shape in geometry, having 4 sides and 4 corners. Its two sides meet at right angles. Thus, a rectangle has 4 angles, each measuring 90 °. The opposite sides of a rectangle have the same lengths and are parallel. Two sides are said to be parallel, when the distance between them remains the same at all points.

Other names of the rectangle:

- Since all the angles of a rectangle are equal, we also call it an equiangular quadrilateral.
- Since it has parallel sides, we can also call it a parallelogram. A parallelogram is a quadrilateral whose opposite sides are equal and parallel.

Properties of a rectangle:

- It is a flat shape.
- It has 4 sides (edges).
- It has 4 corners (vertices).
- It has 4 right angles.

Area of a rectangle: The is the product of its two adjacent sides: Area of a rectangle = Length × Width.

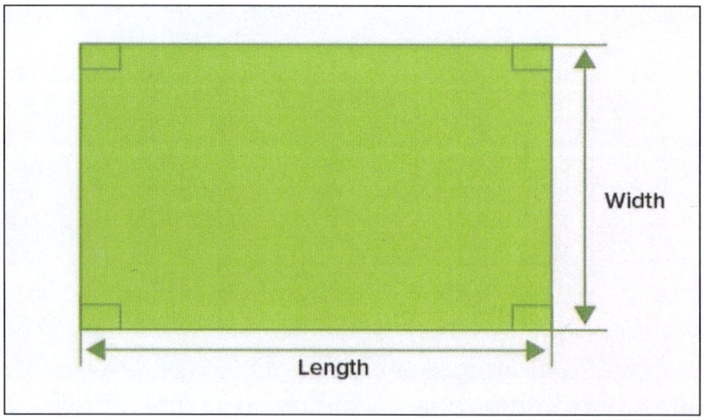

Perimeter

The perimeter of a rectangle is the sum of the length of its four sides. Since parallel sides of a rectangle have the same length, the formula for the perimeter of a rectangle is:

Perimeter of a rectangle = Length + Width + Length + Width

= 2 length + 2Width

= 2 (length + Width)

Diagonals of a Rectangle

The line segments that join the opposite corners of a rectangle are called its diagonals. In the given figure, the two diagonals of the rectangle are AC and BD. The diagonals of a rectangle are the same in length. Therefore, AC = BD.

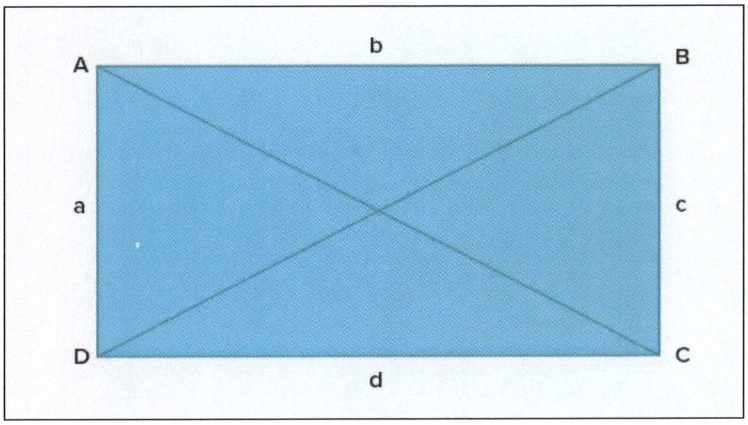

Parallelogram

A parallelogram is a quadrilateral with opposite sides parallel (and therefore opposite angles equal). A quadrilateral with equal sides is called a rhombus and a parallelogram whose angles are all right angles is called a rectangle. And, since a square is a degenerate case of a rectangle, both squares and rectangles are special types of parallelograms.

The polygon diagonals of a parallelogram bisect each other.

The angles of a parallelogram satisfy the identities:

$$A = C$$
$$B = D$$

And,

$$A + B = 180\, degrees.$$

A parallelogram of base b and height h has area:

$$A = b\ h = ab\sin A = ab\sin B.$$

and the polygon diagonals p and q are:

$$p = \sqrt{(a^2 + b^2 - 2ab\cos A)}$$
$$= \sqrt{(a^2 + b^2 + 2ab\cos B)}$$
$$q = \sqrt{(a^2 + b^2 + 2ab\cos A)}$$
$$= \sqrt{(a^2 + b^2 - 2ab\cos B)}$$

The sides a, b, c, d and diagonals p, q of a parallelogram satisfy:

$$p^2 + q^2 = 2(a^2 + b^2)$$

The area of the parallelogram with sides formed by the vectors u = (u_x, u_y) and v = (v_x, v_y) is:

$$A = u \times v$$
$$= det(uv)$$
$$= u_x v_y - u_y v_x,$$

where $u_x v_y$ is the two-dimensional cross product and det A is the determinant.

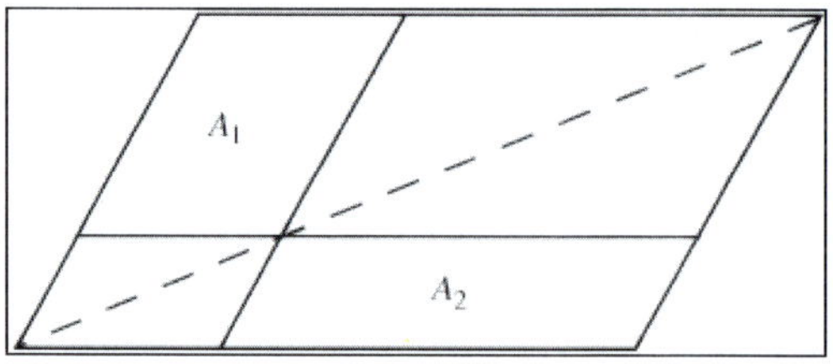

As shown by Euclid, if lines parallel to the sides are drawn through any point on a diagonal of a parallelogram, then the parallelograms not containing segments of that diagonal are equal in area (and conversely), so in the above figure, $A_1 = A_2$.

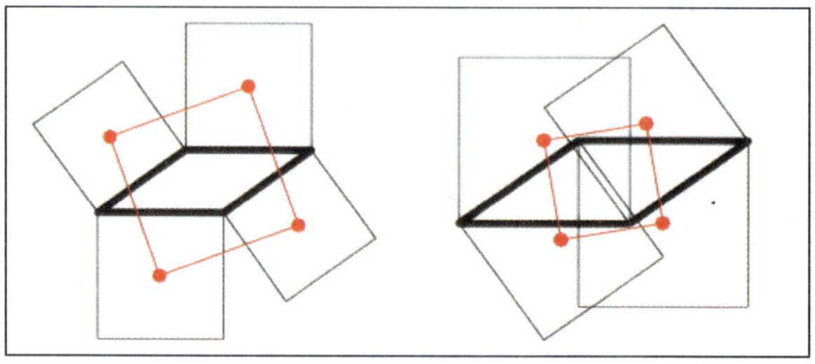

The centers of four squares erected either internally or externally on the sides of a parallelograms are the vertices of a square.

Properties of Parallelogram

Opposite Sides are Parallel

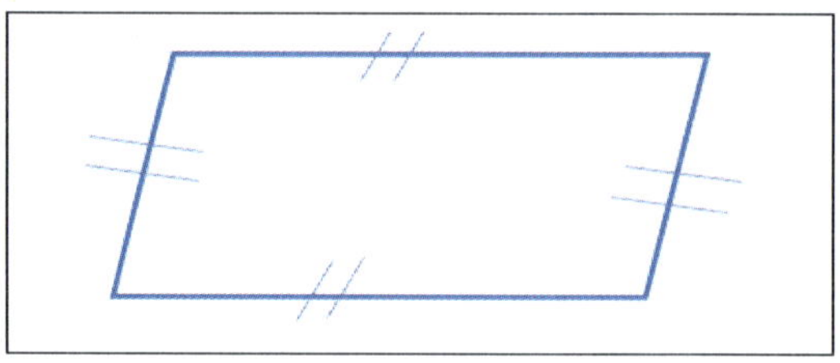

Parallel lines are lines that are always the same distance apart and never touch. If the sides of a parallelogram were lines that continued on, the ones opposite of each other would never touch. These lines would remain the same distance away from each other no matter how far they extended.

Opposite Sides are Congruent

In geometry, congruent means that two things are identical. If you were to superimpose the shapes on top of each other they would match up exactly. This is true for a parallelogram's sides. Each of the opposite sides are the same in length. If you were to break the shape apart and place the opposite sides on top of each other, you would find that they line up perfectly.

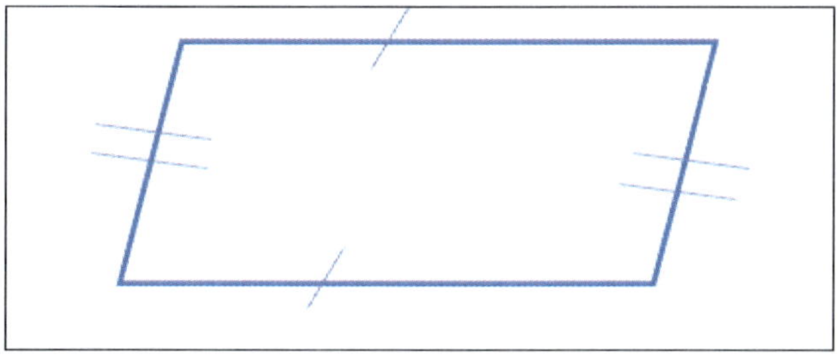

Opposite Angles are Congruent

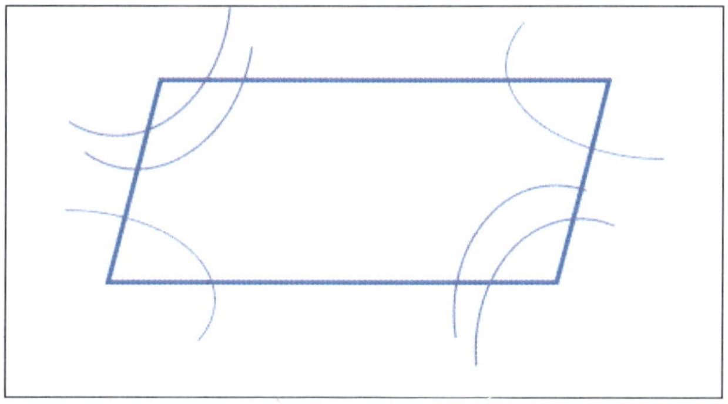

The angles that are opposite of each other are also congruent. To find out if your quadrilateral is a parallelogram, you could get out your protractor and measure each angle. The angles opposite of each other will have the same measurement. It's common for a parallelogram to have two acute angles and two obtuse angles. Therefore, the acute angles should have the same measurement and the obtuse angles should also have the same measurement.

Consecutive Angles are Supplementary

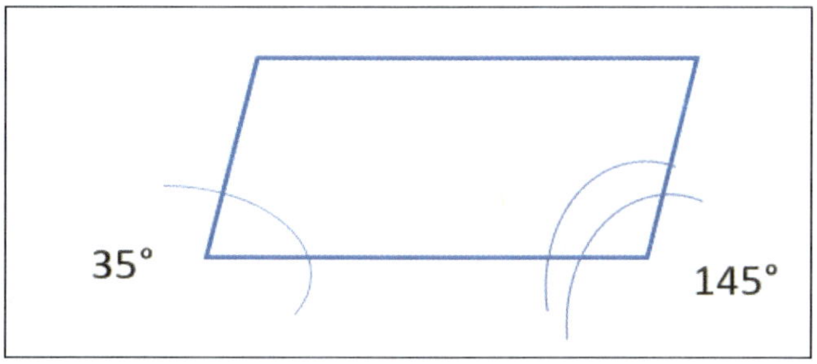

To find another one of the properties of parallelograms, draw an imaginary line through the shape to cut it in half. Then, look at the consecutive angles (or the ones that are next to each other). If the shapes are supplementary, then the shape might be a parallelogram.

Supplementary angles are two angles that add up to 180-degrees. Let's say that two of the consecutive angles have measurements of 35-degrees and 145-degrees. If we add these together (35 + 145), the sum is 180-degrees. Therefore, we have supplementary angles.

Diagonals Bisect each other

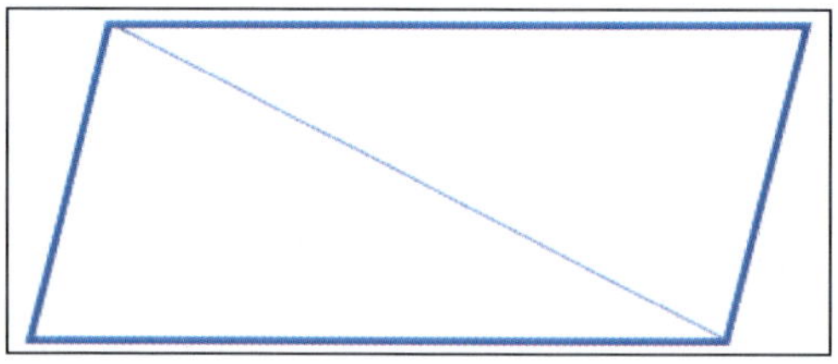

Now pretend to draw an imaginary line from one angle to its opposite, congruent angle. This line should create two congruent triangles within the shape.

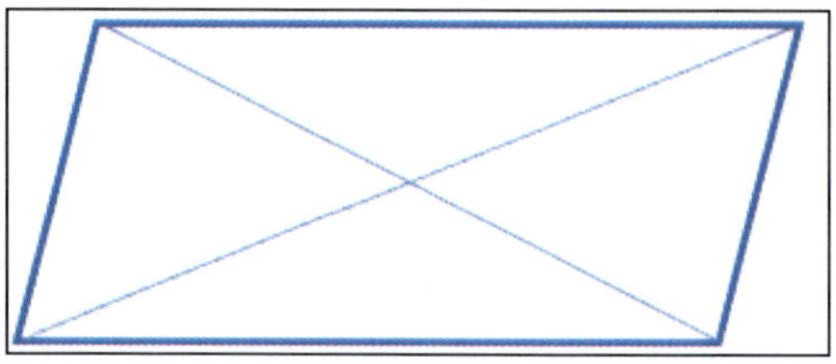

From there, proceed to draw another imaginary line from the supplementary angle to its opposite, congruent angle. These two imaginary lines should bisect one another. (To bisect is to cut something into two equal parts).

If One Angle is a Right Angle

The last property only matters if there is a right angle in your quadrilateral. If you have one angle that is a right angle, then all the rest of the angles should be right angles, too. Why? Because we know that the opposite angles are congruent. We also know that consecutive angles are supplementary and 90 + 90 = 180. Therefore, all four angles would have a measurement of 90-degrees.

Quadrilateral is a parallelogram if it has these properties of parallelograms:

- The opposite sides are parallel.
- The opposite sides are congruent.
- The opposite angles are congruent.
- Consecutive angles are supplementary (add up to 180-degrees).
- The diagonals bisect each other.
- And all four angles measure 90-degrees IF one angle measures 90-degrees.

Trapezoid

A quadrilaterals which have two parallel sides. It is also called a Trapezium in some parts of the world. A trapezoid is a four-sided closed shape or figure which cover some area and also has its perimeter. It is a 2D figure and not 3D figure. The sides which are parallel to each other are termed as the bases of the trapezoid. The non-parallel sides are known as legs or lateral sides. The distance between the parallel sides is known as the altitude.

There is some disagreement over the definition of trapezoids. One school of mathematics considers that a trapezoid can have one and only one pair of parallel sides, while the other argues that there can be more than one pair of parallel sides in a trapezoid. If we consider the second definition, a parallelogram is also a trapezoid according to that. But the first definition does not consider a parallelogram to be a trapezoid. Since we have already mentioned it one of the types of quadrilaterals.

Types of Trapezoids

Trapezoids can be broadly classified into three groups:

- Right Trapezoids,
- Isosceles Trapezoids,
- Scalene Trapezoids.

Right Trapezoids

- A Right trapezoid has a pair of right angles.

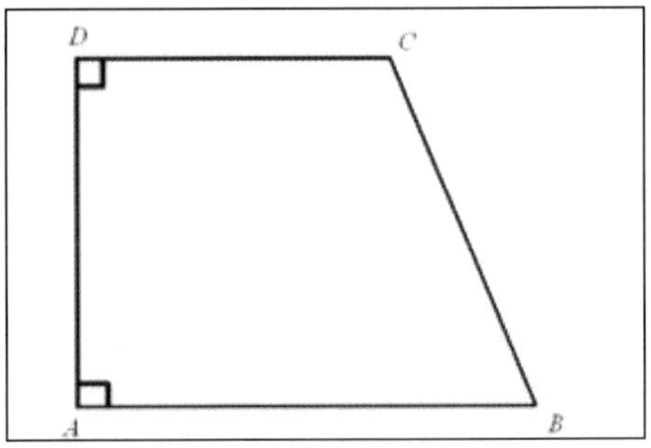

Isosceles Trapezoids

- If the non-parallel sides or the legs of a trapezoid are equal in length, it is known as an Isosceles trapezoid.

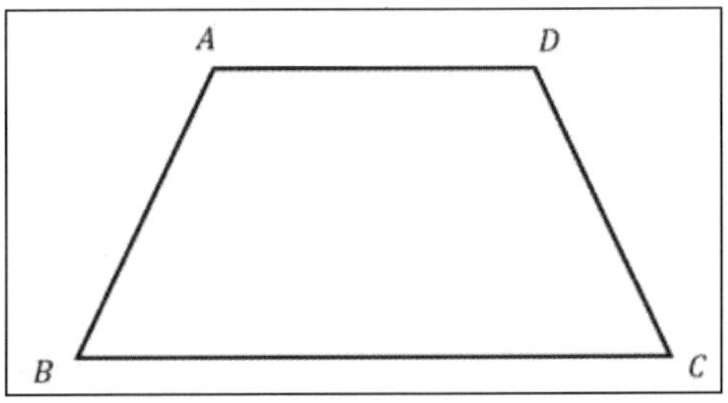

Scalene Trapezoids

- When neither the sides nor the angles of a trapezium are equal, we call it a Scalene trapezoid.

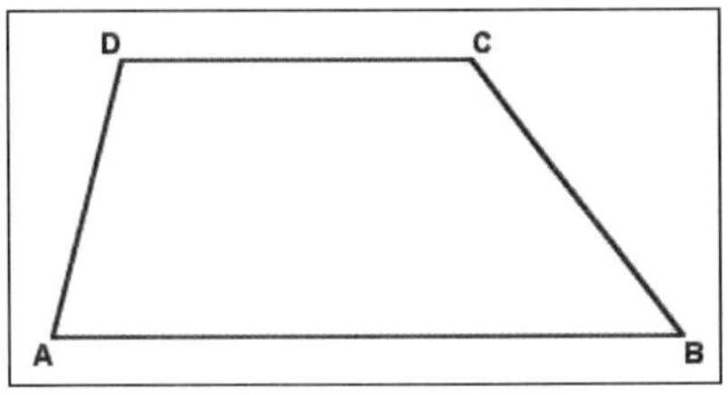

Area of Trapezoid

A trapezoid is a quadrilateral with two sides parallel. The trapezoid is equivalent to the British definition of trapezium. An isosceles trapezoid is a trapezoid in which the base angles are equal so c=d. A right trapezoid is a trapezoid having two right angles.

The area of the trapezoid is:

$$\begin{aligned} A &= \frac{1}{2}(a+b)h \\ &= mh \\ &= \frac{1}{4}\frac{(b+a)}{(b-a)}\eta. \end{aligned}$$

The geometric centroid lies on the median m between the base and top, if the lower left-hand corner of the trapezoid is at the original, lies at:

$$\bar{x} = \frac{b}{2} + \frac{(2a+b)(c^2-d^2)}{6(b^2-a^2)}$$

$$\bar{y} = \frac{b+2a}{3(a+b)}h$$

$$= \frac{b+2a}{6(b^2-a^2)}\eta$$

The trapezoid depicted has central median:

$$m = \frac{1}{2}(a+b)$$

If vertical lines are extended from the endpoints of the upper side, the bases of the triangles formed on the left and right are:

$$x_1 = \frac{(b-a)^2 + c^2 - d^2)}{2(b-a)}$$

$$x_2 = \frac{(b-a)^2 - c^2 + d^2)}{2(b-a)}$$

respectively. This gives the vertex angles as:

$$\theta_1 = \cos^{-1}\left(\frac{(b-a)^2 - c^2 + d^2)}{2(b-a)c}\right)$$

$$\theta_2 = \cos^{-1}\left(\frac{(b-a)^2 - c^2 + d^2)}{2(b-a)c}\right)$$

$$\theta_3 = \cos^{-1}\left(\frac{(a-b)^2 - c^2 + d^2)}{2(a-b)c}\right)$$

$$\theta_4 = \cos^{-1}\left(\frac{(a-b)^2 - c^2 + d^2)}{2(a-b)c}\right)$$

from the lower left corner proceeding counterclockwise.

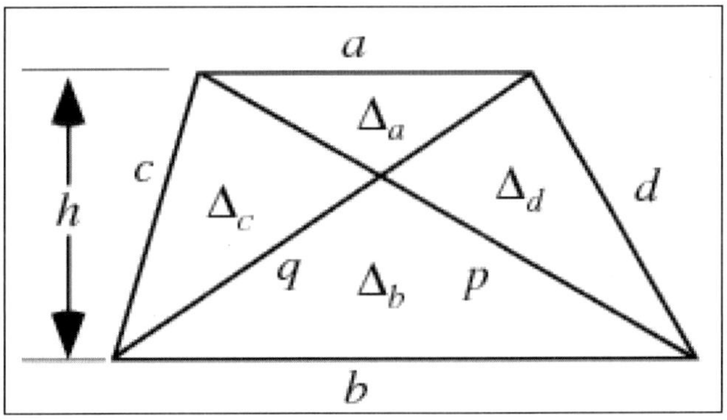

In terms of the side length, the diagonals of the trapezoid are given by:

$$p = \sqrt{\frac{ab^2 - a^2b - ac^2 + bd^2}{b-a}}$$

$$q = \sqrt{\frac{ab^2 - a^2b - ad^2 + bc^2}{b-a}}$$

and the height by:

$$h = \frac{\eta}{2|b-a|}$$

Where,

$$\eta = \sqrt{(-a+b+c+d)(a-b+c+d)(a-b+c-d)(a-b-c+d)}$$
$$= \sqrt[4]{(s-a)(s-b)(s-b-c)(s-b-d)}$$

Where,

$$s = \frac{1}{2}(a+b+c+d)$$

is the semiperimeter.

Letting the lower left vertex be located at the origin, the intersection of the diagonals occurs at:

$$x_0 = \frac{b(a^2 - b^2 - c^2 + d^2)}{2(a^2 - b^2)}$$

$$y_0 = \frac{b\eta}{2(a^2 - b^2)}$$

The areas of the indicated triangles are:

$$\Delta_a = \frac{a^2}{2(a+b)}h$$

$$\Delta_b = \frac{b^2}{2(a+b)}h$$

$$\Delta_c = \frac{ab}{2(a+b)}h$$

$$\Delta_d = \frac{ab}{2(a+b)}h$$

So, $\Delta_c = \Delta_d$

And, $\Delta_a \Delta_b = \Delta_c \Delta_d = \frac{a^2 b^2}{2(a+b)^2} h^2$

Perimeter of Trapezoid

The perimeter of a trapezoid is the sum of all its sides. Therefore, for a trapezoid with sides a, b, c and d, the formula of the perimeter can be written as:

Perimeter= a+b+c+d

Trapezoid Properties

There are certain properties of trapezoids that identify them as trapezoids:

- The base angles and the diagonals of an isosceles trapezoid are equal.
- If you draw a median on a trapezoid, it will be parallel to the bases and its length will be the average of the length of the bases.

- The intersection point of the diagonals is collinear to the midpoints of the two opposite sides.
- If there is a trapezoid with sides a, b, c and d and diagonals p and q, the following equation will be true:

  $p^2+q^2 = c^2+d^2+2ab$.

Example: Find the area of a trapezoid with bases of 3 meters and 5 meters and a height of 4 meters.

Solution: We know that the area of a Trapezoid is $\frac{1}{2}(a+b)\times h$,

$$Area = \frac{3+5}{2}\times 4m^2$$

$$Area = \frac{8}{2}\times 4m^2$$

$$= 16m^2$$

Example: A trapezoid has four sides measuring 3m, 5m, 7m and 4m. Find its perimeter.

Solution: We know that Perimeter is given by Sum of all the sides.

Therefore, Perimeter = 3+5+7+4 = 19 m.

Example: Is a parallelogram also a trapezoid?

Solution: No, a trapezoid is not a parallelogram.

As per definition: Parallelogram has two pairs of parallel sides. While Trapezoid have Exactly two parallel sides.

References

- Geometry, science: britannica.com, Retrieved 10 July, 2019
- Points-lines-and-shapes: math-only-math.com, Retrieved 08 June, 2019
- Understanding-angles-types: mathstips.com, Retrieved 25 August, 2019
- Triangles-and-quadrilaterals, guides-maths-basic-geometrical-ideas: toppr.com, Retrieved 09 May, 2019
- Circle: newworldencyclopedia.org, Retrieved 15 August, 2019
- Squares, facts-geometry: kidsmathgamesonline.com, Retrieved 04 March, 2019
- Rectangle, math-vocabulary-geometry: splashlearn.com, Retrieved 14 July, 2019
- Parallelogram: mathworld.wolfram.com, Retrieved 10 February, 2019
- Trapezoids, maths: byjus.com, Retrieved 18 August, 2019

2

Euclidean Geometry: An Integrated Study

Euclidean geometry is the study and analysis of solid shapes, figures and planes, on the basis of Euclid's axioms, postulates and theorems. Euclidean plane geometry, Euclid's postulates, parallelogram law, Apollonius theorem, etc. fall under its domain. This chapter delves into detailed study of the Euclidean geometry to provide in-depth understanding of it.

Euclidean geometry is a mathematical well-known system attributed to the Greek mathematician Euclid of Alexandria. The method consists of assuming a small set of intuitively appealing axioms, and then proving many other propositions (theorems) from those axioms. Euclid was the first to show how these propositions could be fitted together into a comprehensive deductive and logical system.

Axiomatic Approach

Following a precedent set in the *Elements*, Euclidean geometry has been exposited as an axiomatic system, in which all theorems ("true statements") are derived from a finite number of axioms. Near the beginning of the first book of the *Elements*, Euclid gives five postulates (axioms):

1. Any two points can be joined by a straight line.
2. Any straight line segment can be extended indefinitely in a straight line.
3. Given any straight line segment, a circle can be drawn having the segment as radius and one endpoint as centre.
4. All right angles are congruent.
5. Parallel postulate. If two lines intersect a third in such a way that the sum of the inner angles on one side is less than two right angles, then the two lines inevitably must intersect each other on that side if extended far enough.

These axioms invoke the following concepts: point, straight line segment and line, side of a line, circle with radius and centre, right angle, congruence, inner and right angles, sum. The following verbs appear: join, extend, draw, intersect. The circle described in postulate 3 is tacitly unique. Postulates 3 and 5 hold only for plane geometry; in three dimensions, postulate 3 defines a sphere.

Postulate 5 leads to the same geometry as the following statement, known as Playfair's axiom, which also holds only in the plane:

> Through a point not on a given straight line, one and only one line can be drawn that never meets the given line.

Postulates 1, 2, 3, and 5 assert the existence and uniqueness of certain geometric figures, and these assertions are of a constructive nature: that is, we are not only told that certain things exist, but are also given methods for creating them with no more than a compass and an unmarked straightedge. In this sense, Euclidean geometry is more concrete than many modern axiomatic systems such as set theory, which often assert the existence of objects without saying how to construct them, or even assert the existence of objects that probably cannot be constructed within the theory.

The *Elements* also include the following five "common notions":

1. Things that equal the same thing also equal one another.
2. If equals are added to equals, then the wholes are equal.
3. If equals are subtracted from equals, then the remainders are equal.
4. Things that coincide with one another equal one another.
5. The whole is greater than the part.

Euclid also invoked other properties pertaining to magnitudes. 1 is the only part of the underlying logic that Euclid explicitly articulated. 2 and 3 are "arithmetical" principles; note that the meanings of "add" and "subtract" in this purely geometric context are taken as given. 1 through 4 operationally define equality, which can also be taken as part of the underlying logic or as an equivalence relation requiring, like "coincide," careful prior definition. 5 is a principle of mereology. "Whole," "part," and "remainder" beg for precise definitions.

In the 19th century, it was realized that Euclid's ten axioms and common notions do not suffice to prove all of theorems stated in the *Elements*. For example, Euclid assumed implicitly that any line contains at least two points, but this assumption cannot be proved from the other axioms, and therefore needs to be an axiom itself. The very first geometric proof in the *Elements,* shown in the figure on the right, is that any line segment is part of a triangle; Euclid constructs this in the usual way, by drawing circles around both endpoints and taking their intersection as the third vertex. His axioms, however, do not guarantee that the circles actually intersect, because they are consistent with discrete, rather than continuous, space. Starting with Moritz Pasch in 1882, many improved axiom systems for geometry have been proposed, the best known being those of Hilbert, George Birkhoff, and Tarski.

The Parallel Postulate

To the ancients, the parallel postulate seemed less obvious than the others; verifying it physically would require us to inspect two lines to check that they never intersected, even at some very distant point, and this inspection could potentially take an infinite amount of time. Euclid himself seems to have considered it as being qualitatively different from the others, as evidenced by the organization of the *Elements*: the first 28 propositions he presents are those that can be proved without it.

Treatment using Analytic Geometry

The development of analytic geometry provided an alternative method for formalizing geometry. In this approach, a point is represented by its Cartesian (x,y) coordinates, a line is represented by its equation, and so on. In the 20th century, this fit into David Hilbert's program of reducing all of mathematics to arithmetic, and then proving the consistency of arithmetic using finitistic reasoning. In Euclid's original approach, the Pythagorean theorem follows from Euclid's axioms. In the Cartesian approach, the axioms are the axioms of algebra, and the equation expressing the Pythagorean theorem is then a definition of one of the terms in Euclid's axioms, which are now considered to be theorems. The equation,

$$|PQ| = \sqrt{(p-r)^2 + (q-s)^2}$$

defining the distance between two points P = (p, q) and Q = (r, s) is then known as the Euclidean metric, and other metrics define non-Euclidean geometries.

Euclid's Postulates

Any statement that is assumed to be true on the basis of reasoning or discussion is a postulate or axiom. The postulates stated by Euclid are the foundation of Geometry and are rather simple observations in nature. 'Euclid' was a Greek mathematician regarded as the 'Father of Modern Geometry'.

Postulate – I

A straight line segment can be formed by joining any two points in space.

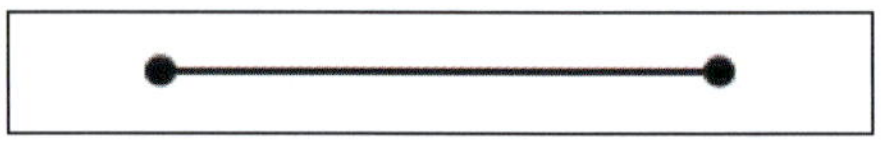

In Geometry, a line segment is a part of a line that is bounded by 2 distinct points on either end. It consists of a series of points bounded by the two endpoints. Thus a line segment is measurable as the distance between the two endpoints. A line segment is named after the two endpoints with an overbar on them.

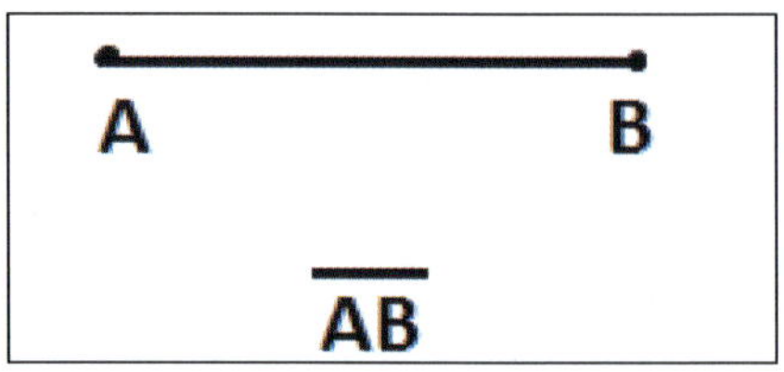

Postulate – II

Any straight line can be extended indefinitely on both sides. Unlike a line segment, a line is not bounded by any endpoint and so can be extended indefinitely in either direction. A line is uniquely defined as passing through two points which are used to name it.

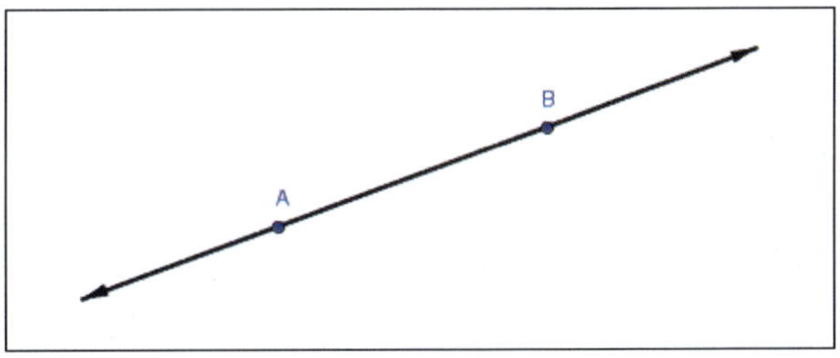

Postulate – III

A circle can be drawn with any centre and any radius. For any line segment, a circle can be drawn with its centre at one endpoint and the radius of the circle as the length of the line segment. Consider a line segment bounded by two points. If one of these points is taken as the centre of a circle and the radius of the circle is taken as equal to the length of the segment, a circle can be drawn with its diameter twice than the length of the line segment.

In the above example, the line segment AO serves as the radius of a circle with centre at point O and a diameter equal to AB where l(AB) =2l(AO).

Postulate – IV

All right angles are congruent or equal to one another. A right angle is an angle measuring 90 degrees. So, irrespective of the length of a right angle or its orientation all right angles are identical in form and coincide exactly when placed one on top of the other.

A right angle.

Postulate – V

Two lines are parallel to each other if they intersect the third line and the interior angle between them is 180 degrees.

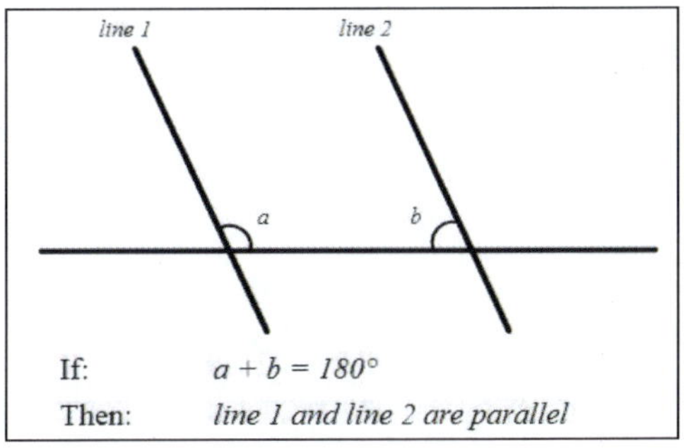

'Parallel lines' are a set of 2 or more lines that never cross or intersect each other at any point in space if they are extended indefinitely. As you can see in the above image, line 1 and line 2 are parallel if and only if the sum of angles 'a' and 'b' they make with the transversal is 180 degrees.

Euclidean Plane

The Euclidean plane is the plane that is the object of study in Euclidean geometry (high-school geometry). The plane and the geometry are named after the ancient-Greek mathematician Euclid.

The Euclidean plane is a collection of points P, Q, R between which a distance ρ is defined, with the properties,

1. $\rho(\mathrm{P},\mathrm{Q}) \in \mathbb{R}$ (real line).
2. $\rho(\mathrm{P},\mathrm{Q}) \geq 0$ and $\rho(\mathrm{P},\mathrm{Q}) = 0$ if and only if P = Q.
3. $\rho(\mathrm{P},\mathrm{Q}) = \rho(\mathrm{Q},\mathrm{P})$.
4. $\rho(\mathrm{P},\mathrm{Q}) \leq \rho(\mathrm{P},\mathrm{R}) + \rho(\mathrm{R},\mathrm{Q})$ (triangle inequality).

In other words, the Euclidean plane is a metric space.

Well-known subsets of the Euclidean plane are straight lines. Any two non-coinciding points P and Q [ρ(P,Q) ≠ 0] in the plane lie on a unique straight line, that is to say, P and Q determine a unique straight line. The length of the straight line segment connecting the points P and Q is by definition the Euclidean distance rho;(P,Q). In a Euclidean plane no upper bound exists on the distance between two points: given a fixed point P and a "running" point Q, the distance ρ(P,Q) is unbounded, so that Q can "run to infinity" without leaving the plane. The straight line segment connecting P and Q may be extended to a half-line, which is a line segment terminated in P and unbounded on the other side. If P also runs to infinity (away from Q), i.e., the straight segment is unbounded on both sides, the segment becomes a straight line. Hence, lines and half-lines in a Euclidean plane are of infinite length.

Straight lines may be intersecting. Two straight lines that do not intersect are called parallel. The point of intersection of two non-parallel straight lines is a single unique point. Angles between intersecting lines can be measured and expressed in degrees. In particular, straight lines may be perpendicular, that is, their angle of intersection is 90°. The maximum number of perpendicular lines that can intersect at a point of a Euclidean plane is equal for all points; this maximum is two for a Euclidean plane. This fact is expressed by the statement that a Euclidean plane is two-dimensional.

A Cartesian coordinate frame can be erected anywhere in the plane. Briefly, one chooses an arbitrary point O (the origin) and let it be the point of intersection of two perpendicular straight lines, one line called the x-axis, the other line called the y-axis. Given an arbitrary point P in the plane, it is possible to construct a half-line from P intersecting the x-axis perpendicularly at the point X: one says that X is the perpendicular projection of P on the x-axis. The non-negative real number x := ρ(O,X) is the absolute value of the x-coordinate of P. The non-negative number y := ρ(P,X) is the absolute value of the y-coordinate of P. It is useful and common to give signs to coordinates. If

Y is the perpendicular projection of P on the y-axis, one sees easily that the distance ρ(Y,P) is equal to |x| and ρ(O,Y) is equal to |y|.

This construction by means of a Cartesian frame (also known as system of axes) shows that in a Euclidean plane every point in the plane is given by a pair of numbers (x, y) defined with respect to one and the same frame. Let P = (x, y) (this is a short-hand notation for P having coordinates x and y) and let Q = (x′, y′). From the Pythagorean theorem (which holds for Euclidean planes) it follows directly that,

$$\rho(P,Q) = \sqrt{(x-x')^2 + (y-y')^2}$$

Suppose the pairs (x, y) and (x", y") with x ≠ x" and y ≠ y" would represent P (both pairs with respect to the same Cartesian frame). Then the distance,

$$\sqrt{(x-x'')^2 + (y-y'')^2} \neq 0$$

and because axiom 2 states that two points with non-zero distance are necessarily distinct, it follows that both pairs of numbers cannot simultaneously represent the same point P. The coordinate pair (x, y) represents P uniquely. In summary,

if $P = (x, y)$ and $Q = (x', y')$ then $P = Q$ if and only if $x = x'$ and $y = y'$.

Different geometric figures (triangles, squares, etc.) can be constructed. A geometric figure can be translated and rotated without change of shape. Such a map is called a rigid motion of the figure. The totality of rigid motions form a group of infinite order, the Euclidean group in two dimensions, often written as E(2). Formally, the Euclidean plane is a 2-dimensional affine space with inner product.

Plane Curve

A plane curve can be expressed in the parametric form as,

$$x = x(t), \quad y = y(t),$$

Where the coordinates of the point (x, y) of the curve are expressed as functions of a parameter t within a closed interval $t_1 \leq t \leq t_2$. The functions $x(t)$ and $y(t)$ are assumed to be continuous with a sufficient number of continuous derivatives. The parametric curve is said to be of class r, if the functions have continuous derivatives up to the order r, inclusively. In vector notation the parametric curve can be specified by a vector-valued function,

$$r = r(t).$$

Another method of representing a curve analytically is to impose one condition on a variable point (x, y) by an equation of the form,

$$f(x, y) = 0.$$

This is an implicit equation for a plane curve. When $f(x,y)$ is linear in variables x and y, $f(x,y)=0$ represents a straight line. If $f(x,y)$ is of the second degree in x and y (i.e. $ax^2+2bxy+cy^2+2dx+2ey+h=0$), $f(x,y)=0$ represents a variety of plane curves called conic sections. The implicit equation for a plane curve can also be expressed as an intersection curve between a parametric surface and a plane.

The explicit form can be considered as a special case of parametric and implicit forms. If t can be expressed as a function of x or y, we can easily eliminate *t* from $x=x(t),\ y=y(t)$, to generate the explicit form,

$$=y=F(x)\ or\ x=G(y)$$

This is always possible at least locally when $\frac{dx}{dt}\neq 0$ or $\frac{dy}{dt}\neq 0$. Conversely if we set *x* or *y* in $=y=F(x)\ or\ x=G(y)$ to be equal to the parameter t we obtain the parametric form $x=x(t),\ y=y(t)$. Also if the implicit equation $f(x,y)=0$ can be solved for one variable in terms of the other, we also obtain $=y=F(x)\ or\ x=G(y)$. This is always possible at least locally when $\frac{\partial f}{\partial y}\neq 0$ or $\frac{\partial f}{\partial x}\neq 0$.

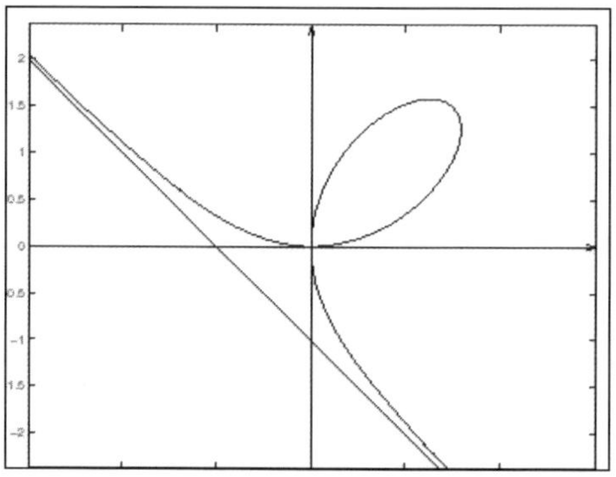

Folium of Descartes.

Figure shows the Folium of Descartes, introduced by R. Descartes in 1638, with its asymptotic line. It can be expressed in parametric form,

$$r(t)=\left(\frac{3t}{1+t^3},\frac{3t^2}{1+t^3}\right)^T,-\infty<t<\infty(t\neq -1),$$

Where superscript *T* denotes transpose of a vector. For $t<-1$ the curve is located in the fourth quadrant and approaches the origin as *t* goes to $-\infty$. For $-1<t<0$ the curve is located in the second quadrant, and $t=0$ corresponds to the origin. In the first quadrant it forms a loop moving counter-clockwise as *t* increases from 0 to $+\infty$ Eliminating *t* from $r(t)=\left(\frac{3t}{1+t^3},\frac{3t^2}{1+t^3}\right)^T,-\infty<t<\infty(t\neq -1)$, the Folium of Decartes can be also expressed in an implicit form,

$$f(x,y)=x^3+y^3-3xy=0.$$

We can easily trace the curve using the parametric equation $r(t)=\left(\frac{3t}{1+t^3},\frac{3t^2}{1+t^3}\right)^T$ $-\infty<t<\infty\,(t\neq -1)$, by evaluating $x(t)$ and $y(t)$ for a discrete sampling of *t*, while such tracing is more difficult when using the implicit equation $f(x,y)=x^3+y^3-3xy=0$. However, determining if a point (x_0,y_0) lies on the curve is easier when using the implicit rather than the parametric equation of the curve. For example, we can verify that the point $\left(\frac{3}{2},\frac{3}{2}\right)$ lies on the curve by substituting $x=\frac{3}{2}$ and $y=\frac{3}{2}$ into implicit form and deducing that $f\left(\frac{3}{2},\frac{3}{2}\right)=0$. However, it is more complex to deduce this using the parametric form. We first set $x(t)=\frac{3}{2}$ which yields a cubic equation $t^3-2t+1=0$. The roots of the cubic equation are 1, $\frac{-1\pm\sqrt{5}}{2}$. Then we substitute each root into $y(t)$ to see if it becomes equal to $\frac{3}{2}$.

Curvature of Plane Curves

Let a plane curve C be defined parametrically by the radius vector **r** (t). While a point M moves along the curve C, the direction of the tangent changes.

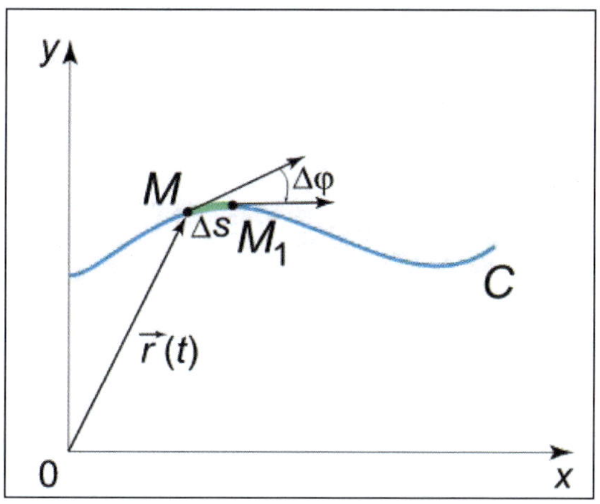

The curvature of the curve can be defined as the ratio of the rotation angle of the tangent Δφ to the traversed arc length $\Delta s = MM_1$. This ratio $\frac{\Delta\varphi}{\Delta s}$ is called the average curvature of the curve. When the point M_1 approaches the point M, we obtain the curvature of the curve at the point M:

$$k=\lim_{\Delta s\to 0}\frac{\Delta\varphi}{\Delta s}=\frac{d\varphi}{ds}.$$

It is clear that the curvature k in the general case can be either positive or negative, depending on the direction of rotation of the tangent.

If a curve is defined by the radius vector **r** (t), its curvature is given by,

$$k = \frac{r' \times r''}{|r'|^3},$$

where ***r′***, ***r′′*** are the first and second derivatives of the radius vector.

In this formula, the numerator contains the vector product of the vectors ***r′*** and ***r′′***.

If the coordinates of a curve are specified parametrically as x(t) and y(t), the formula for calculating the curvature takes the form,

$$k = \frac{x'y'' - y'x''}{\left[(x')^2 + (y')^2\right]^{\frac{3}{2}}}.$$

If a plane curve is given by an explicit function $y = f(x)$, the curvature is calculated by the formula,

$$k = \frac{y''}{\left[1 + (y')^2\right]^{\frac{3}{2}}}.$$

In the case when a curve is given in polar coordinates in the form $\rho = \rho(\varphi)$, its curvature k is defined by the expression,

$$k(\varphi) = \frac{\rho^2 + 2(\rho')^2 - \rho\rho''}{\left[\rho^2 + (\rho')^2\right]^{\frac{3}{2}}}.$$

The curvature of the curve is often understood as the absolute value of curvature, without taking into account the direction of rotation of the tangent. In this case, the above formulas remain valid, but the absolute value appears in the numerator. For example, the formula for the curvature when the coordinates *x(t)* and *y(t)* of a curve are given parametrically will look as follows:

$$k = \frac{|x'y'' - y'x''|}{\left[(x')^2 + (y')^2\right]^{\frac{3}{2}}}.$$

The reciprocal of the curvature, is called the radius of curvature:

$$R = \frac{1}{|k|}.$$

The circle with this radius and the center, located on the inner normal line, will most closely approximate the plane curve at the given point.

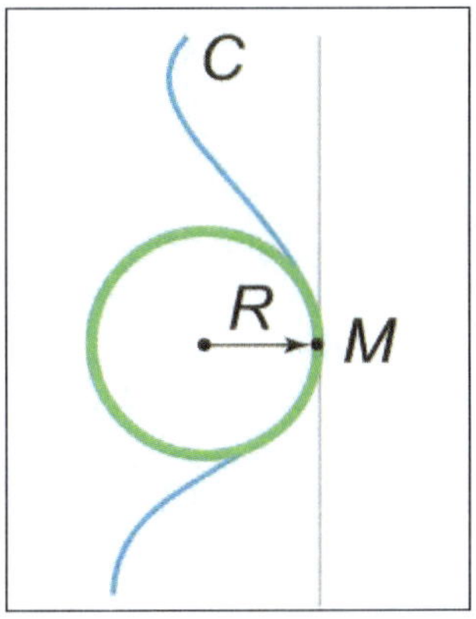

Such a circle is called the osculating circle.

Parallelogram Law

The parallelogram law is the fundamental law that belongs to elementary Geometry. This law is also known as parallelogram identity.

Parallelogram Law of Addition

Parallelogram law states that the sum of the squares of the length of the four sides of a parallelogram is equal to the sum of the squares of the length of the two diagonals. In Euclidean geometry, it is necessary that the parallelogram should have equal opposite sides.

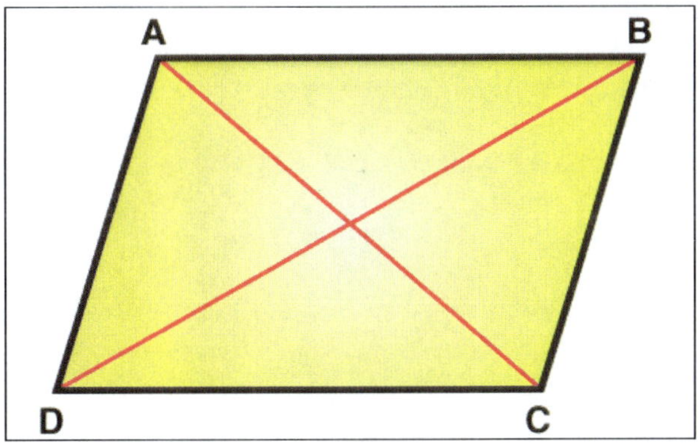

If ABCD is a parallelogram, then AB = DC and AD = BC. Then according to the definition of the parallelogram law, it is stated as:

$$2(AB)^2 + 2(BC)^2 = (AC)^2 + (BD)^2$$

In case the parallelogram is a rectangle, then the law is stated as:

$$2(AB)^2 + 2(BC)^2 = 2(AC)^2$$

Because in rectangle, two diagonals are of equal lengths. i.e., (AC=BD).

Parallelogram Law of Vectors

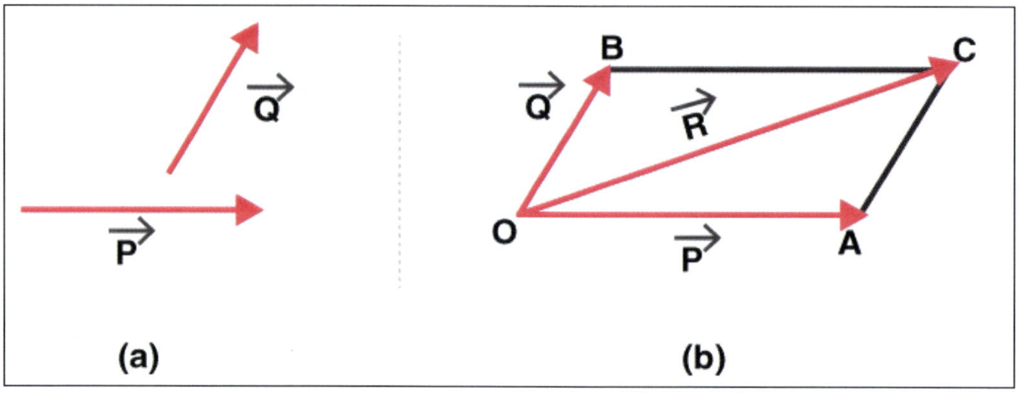

If two vectors are acting simultaneously at a point, then it can be represented both in magnitude and direction by the adjacent sides drawn from a point. Therefore, the resultant vector is completely represented both in direction and magnitude by the diagonal of the parallelogram passing through the point.

The vector P and vector Q represents the sides, OA and OB, respectively. According to the parallelogram law, the side OC of the parallelogram represents the resultant vector R,

$$\overrightarrow{OA} + \overrightarrow{OB} = \overrightarrow{OC} \text{ or}$$

$$\vec{P} + \vec{Q} = \vec{R}$$

Procedure of the Parallelogram Law of Addition of Vectors

The steps for the parallelogram law of addition of vectors are:

- Draw a vector using a suitable scale in the direction of the vector.
- Draw the second vector using the same scale from the tail of the first vector.
- Treat these vectors as the adjacent sides and complete the parallelogram.
- Now, the diagonal represents the resultant vector in both magnitude and direction.

Parallelogram Law Proof

Let AD = BC = x, AB = DC = y, and $\angle$ BAD = α

Using the law of cosines in the triangle BAD, we get,

$$x^2 + y^2 - 2xy\cos(\alpha) = BD^2$$

We know that in a parallelogram, the adjacent angles are supplementary. So,

$$\angle ADC = 180 - \alpha$$

Now, again use the law of cosines in the triangle ADC,

$$x^2 + y^2 - 2xy\cos(180 - \alpha) = AC^2$$

Apply trigonometric identity cos(180 – x) = – cos x in $x^2 + y^2 - 2xy\cos(180 - \alpha) = AC^2$,

$$x^2 + y^2 + 2xy\cos(\alpha) = AC^2$$

Now, the sum of the squares of the diagonals $(BD^2 + AC^2)$ are represented as,

$$BD^2 + AC^2 = x^2 + y^2 - 2xy\cos(\alpha) + x^2 + y^2 + 2xy\cos(\alpha)$$

Simplify the above expression, we get,

$$BD^2 + AC^2 = 2x^2 + 2y^2$$

The above equation is represented as:

$$BD^2 + AC^2 = 2(AB)^2 + 2(BC)^2$$

Hence, the parallelogram law is proved.

Apollonius Theorem

In any triangle, the sum of squares of any two sides is equal to the sum of half the square of third side and twice the square of corresponding mediani.e. if a, b, c are three sides BC, AC, and AB respectively & m is the length of median in a triangle ΔABC (as shown in the figure) then,

$$b^2 + c^2 = \frac{a^2}{2} + 2m^2$$

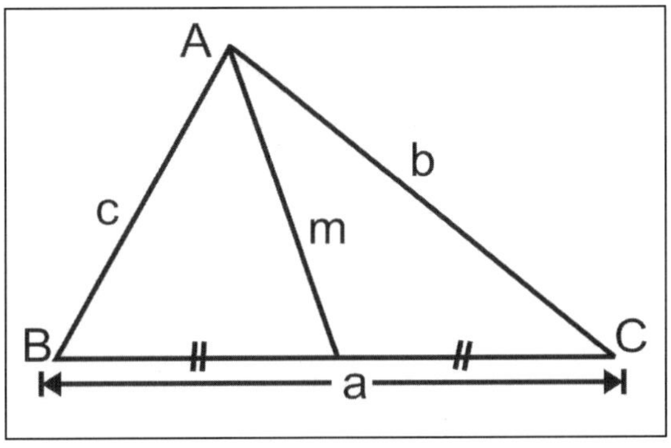

ΔABC having sides *a*, *b*, *c*, & a median of length *m*.

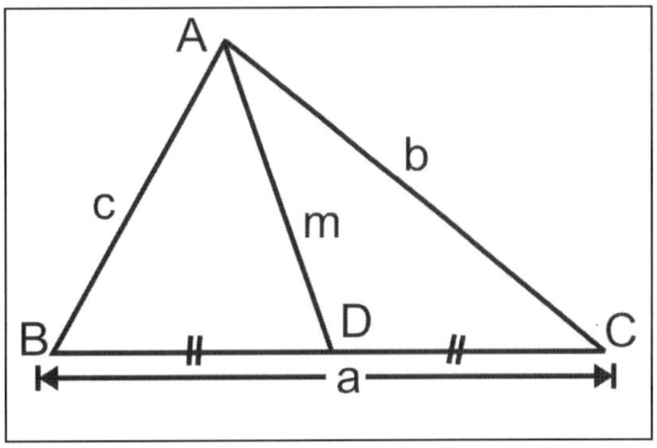

ΔABC, $BD = CD = a/2$ & $AD = m$.

Proof: (Using Trigonometry): Consider a triangle ΔABC having sides BC, AC and AB of lengths a, b, & c respectively and the median AD of length m (as shown in the figure).

Now, applying Cosine rule in ΔABD:

$$\cos B = \frac{AB^2 + BD^2 - AD^2}{2(AB)(BD)}$$

$$\cos B = \frac{c^2 + \left(\frac{a}{2}\right)^2 - m^2}{2(c)\left(\frac{a}{2}\right)}$$

$$ac\cos B = c^2 + \frac{a^2}{4} - m^2.$$

Again, applying Cosine rule in ΔACD:

$$\cos C = \frac{AC^2 + CD^2 - AD^2}{2(AC)(CD)}$$

$$\cos C = \frac{b^2 + \left(\frac{a}{2}\right)^2 - m^2}{2(b)\left(\frac{a}{2}\right)}$$

$$ab\cos C = b^2 + \frac{a^2}{4} - m^2$$

Adding, $ac\cos B = c^2 + \frac{a^2}{4} - m^2$ & $ab\cos C = b^2 + \frac{a^2}{4} - m^2$, we get,

$$ac\cos B + ab\cos C = c^2 + \frac{a^2}{4} - m^2 + b^2 + \frac{a^2}{4} - m^2$$

$$a(b\cos C + c\cos B) = \frac{a^2}{2} + b^2 + c^2 - 2m^2$$

Setting b = $K\sin B$ & c = $K\sin C$ from Sine rule in ΔABC:

$$a(K\sin B\cos C + K\sin C\cos B) = \frac{a^2}{2} + b^2 + c^2 - 2m^2$$

$$aK(\sin B\cos C + \sin C\cos B) = \frac{a^2}{2} + b^2 + c^2 - 2m^2$$

$$aK\sin(B + C) = \frac{a^2}{2} + b^2 + c^2 - 2m^2$$

$$aK\sin(\pi - A) = \frac{a^2}{2} + b^2 + c^2 - 2m^2 \qquad (\text{since, } A + B + C = \pi)$$

$$aK\sin A = \frac{a^2}{2} + b^2 + c^2 - 2m^2$$

$$a \cdot a = \frac{a^2}{2} + b^2 + c^2 - 2m^2 \qquad (\textit{from Sine rule}, a = K\sin A)$$

$$a^2 = \frac{a^2}{2} + b^2 + c^2 - 2m^2$$

$$b^2 + c^2 = \frac{a^2}{2} + 2m^2 \qquad \text{Proved}$$

Proof: (Using Trigonometry): Consider a triangle ΔABC having sides *BC*, *AC* and *AB* of lengths *a, b, & c* respectively and the median AD of length. Let $\angle BAD = \alpha, \angle CAD = \beta$ & $\angle ADB = \theta$ (as shown in the figure).

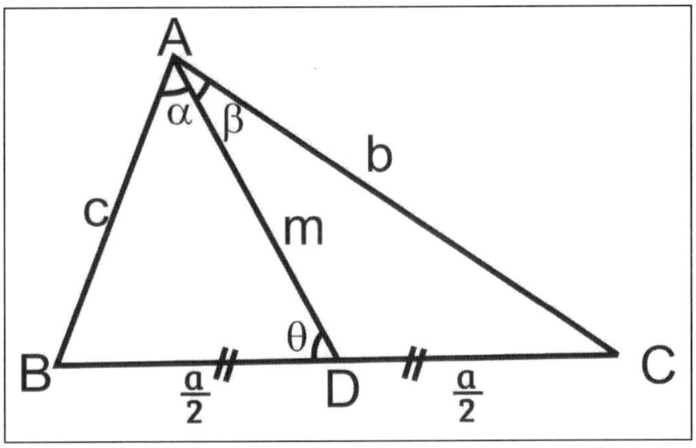

In $\Delta ABC, \angle BAD = \alpha, \angle CAD = \beta$ & $\angle ADB = \theta$.

Applying sine rule in ΔABD:

$$\frac{\sin\alpha}{a/2} = \frac{\sin\theta}{c}$$

$$\sin\theta = \frac{2c}{a}\sin\alpha$$

Applying sine rule in ΔACD:

$$\frac{\sin\beta}{a/2} = \frac{\sin(\pi - \theta)}{b}$$

$$\sin\theta = \frac{2b}{a}\sin\beta$$

Equating values of $\sin\theta$ from $\sin\theta = \frac{2c}{a}\sin\alpha$ & $\sin\theta = \frac{2b}{a}\sin\beta$, we get,

$$\frac{2c}{a}\sin\alpha = \frac{2b}{a}\sin\beta$$

$$\sin\beta = \frac{c}{b}\sin\alpha$$

Applying cosine rule in ΔABD:

$$\cos\alpha = \frac{m^2 + c^2 - \left(\frac{a}{2}\right)^2}{2mc} = \frac{4m^2 + 4c^2 - a^2}{8mc}$$

Applying cosine rule in ΔACD:

$$\cos\beta = \frac{m^2 + b^2 - \left(\frac{a}{2}\right)^2}{2mb} = \frac{4m^2 + 4b^2 - a^2}{8mb}$$

Applying cosine rule in ΔABC:

$$\cos(\alpha + \beta) = \frac{b^2 + c^2 - a^2}{2bc}$$

$$\cos\alpha\cos\beta - \sin\alpha\sin\beta = \frac{b^2 + c^2 - a^2}{2bc}$$

Setting the values of $\sin\beta, \cos\alpha$ & $\cos\beta$ from $\sin\beta = \frac{c}{b}\sin\alpha$, $\cos\alpha = \frac{m^2 + c^2 - \left(\frac{a}{2}\right)^2}{2mc}$ $= \frac{4m^2 + 4c^2 - a^2}{8mc}$ & $\cos\beta = \frac{m^2 + b^2 - \left(\frac{a}{2}\right)^2}{2mb} = \frac{4m^2 + 4b^2 - a^2}{8mb}$ respectively as follows:

$$\left(\frac{4m^2 + 4c^2 - a^2}{8mc}\right)\left(\frac{4m^2 + 4b^2 - a^2}{8mb}\right) - \sin\alpha\left(\frac{c}{b}\sin\alpha\right) = \frac{b^2 + c^2 - a^2}{2bc}$$

$$\frac{\left(4m^2 + 4c^2 - a^2\right)\left(4m^2 + 4b^2 - a^2\right)}{64m^2bc} - \frac{c}{b}\sin^2\alpha = \frac{b^2 + c^2 - a^2}{2bc}$$

$$\frac{\left(4m^2 - a^2 + 4c^2\right)\left(4m^2 - a^2 + 4b^2\right)}{64m^2bc} - \frac{c}{b}\left(1 - \cos^2\alpha\right) - \frac{b^2 + c^2 - a^2}{2bc} = 0$$

$$\frac{\left(4m^2-a^2\right)^2+4\left(b^2+c^2\right)\left(4m^2-a^2\right)+16b^2c^2}{64m^2bc}-\frac{c}{b}\left(1-\left(\frac{4m^2+4c^2-a^2}{8mc}\right)^2\right)-\frac{b^2+c^2-a^2}{2bc}=0$$

$$\frac{\left(4m^2-a^2\right)^2+4\left(b^2+c^2\right)\left(4m^2-a^2\right)+16b^2c^2}{64m^2bc}-\frac{c}{b}\left(\frac{64m^2c^2-\left(4m^2-a^2\right)^2-16c^4-8c^2\left(4m^2-a^2\right)}{64m^2c^2}\right)$$

$$-\frac{b^2+c^2-a^2}{2bc}=0$$

$$\frac{\left(4m^2-a^2\right)^2+4\left(b^2+c^2\right)\left(4m^2-a^2\right)+16b^2c^2-64m^2c^2+\left(4m^2-a^2\right)^2+16c^4+8c^2\left(4m^2-a^2\right)}{64m^2bc}$$

$$\frac{2\left(4m^2-a^2\right)^2+4\left(b^2+3c^2\right)\left(4m^2-a^2\right)+16b^2c^2-64m^2c^2+16c^4-32m^2\left(b^2+c^2-a^2\right)}{64m^2bc}=0$$

$$16m^4+a^4-8m^2a^2+8m^2b^2+24m^2c^2-2a^2b^2-6a^2c^2+8b^2c^2-32m^2c^2+8c^4-16m^2b^2-16m^2c^2$$
$$+16m^2a^2=0$$

$$(16m^4+a^4+8m^2a^2)-8m^2b^2-24m^2c^2-2a^2b^2-6a^2c^2+8b^2c^2+8c^4=0$$

$$\left(4m^2+a^2\right)^2-8m^2\left(b^2+3c^2\right)-2a^2\left(b^2+3c^2\right)+8c^2\left(b^2+c^2\right)=0$$

$$\left(4m^2+a^2\right)^2-2\left(b^2+3c^2\right)\left(4m^2+a^2\right)+8c^2\left(b^2+c^2\right)=0$$

Let $4m^2+a^2=x$ then we get a quadratic equation as follows,

$$x^2-2\left(b^2+3c^2\right)x+8c^2\left(b^2+c^2\right)=0$$

$$x=\frac{-\left(-2\left(b^2+3c^2\right)\right)\pm\sqrt{\left(-2\left(b^2+3c^2\right)\right)^2-4\times1\times8c^2\left(b^2+c^2\right)}}{2\times1}$$

$$x=b^2+3c^2\pm\sqrt{b^4+9c^4+6b^2c^2-86b^2c^2-8c^4}$$

$$x=b^2+3c^2\pm\sqrt{b^4+c^4-2b^2c^2}$$

$$x=b^2+3c^2\pm\left(b^2-c^2\right)$$

Case-1: Taking positive sign, we get,

$$x=b^2+3c^2+b^2-c^2$$

$$\Rightarrow 4m^2+a^2=2\left(b^2+c^2\right)$$

Case-2: Taking negative sign, we get,

$$x=b^2+3c^2-\left(b^2-c^2\right)$$

$$4m^2+a^2=4c^2$$

This case holds only if ΔABC is an isosceles triangle i.e. for b = c but it is not the case i.e. this case does not meet the requirements of a scalene triangle.

We accept only case-1 which gives us,

$$4m^2 + a^2 = 2\left(b^2 + c^2\right)$$

$$b^2 + c^2 = \frac{1}{2}\left(4m^2 + a^2\right)$$

$$b^2 + c^2 = \frac{a^2}{2} + 2m^2 \qquad \text{Proved}$$

Proof: (Using Pythagoras Theorem): Consider a triangle ΔABC having sides BC, AC and AB of lengths a, b & respectively and the median AD of length m. Drop a perpendicular AN from vertex A to the side BC (as shown by dotted line AN in the figure).

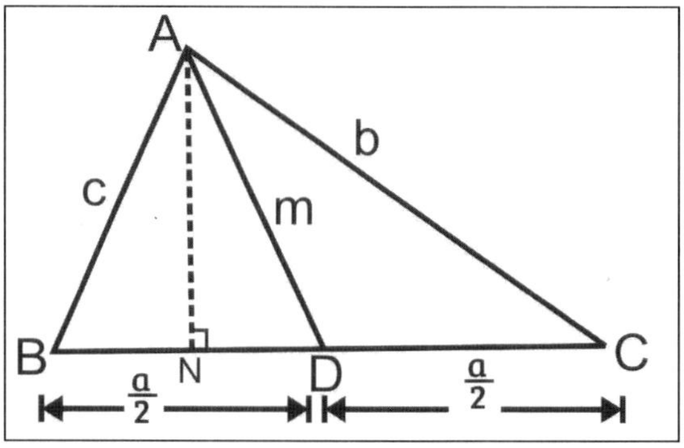

In ΔABC, AD is median of length m & BD = CD = a/2.

Applying Pythagoras Theorem in right ΔAND:

$$AN^2 + ND^2 = AD^2$$

$$AN^2 + \left(\frac{a}{2} - ND\right)^2 = c^2$$

$$AN^2 + ND^2 + \frac{a^2}{4} - a\left(ND\right) = c^2$$

$$m^2 + \frac{a^2}{4} - a\left(ND\right) = c^2 \quad \left(\text{from, } AN^2 + ND^2 = m^2\right)$$

$$a\left(ND\right) = m^2 + \frac{a^2}{4} - c^2$$

Applying Pythagoras Theorem in right ANC:

$$AN^2 + NC^2 = AC^2$$

$$AN^2 + \left(\frac{a}{2} + ND\right)^2 = b^2$$

$$AN^2 + ND^2 + \frac{a^2}{4} + a(ND) = b^2$$

$$m^2 + \frac{a^2}{4} + a(ND) = b^2 \qquad \left(\textit{from},\ AN^2 + ND^2 = m^2\right)$$

$$m^2 + \frac{a^2}{4} + \left(m^2 + \frac{a^2}{4} - c^2\right) = b^2 \qquad \left(\textit{setting value of } a(ND)\ \textit{from}\, a(ND) = m^2 + \frac{a^2}{4} - c^2\right)$$

$$m^2 + \frac{a^2}{4} + m^2\frac{a^2}{4} - c^2 = b^2$$

$$2m^2 + \frac{a^2}{2} = b^2 + c^2$$

$$b^2 + c^2 = \frac{a^2}{2} + 2m^2 \qquad \text{Proved}$$

Apollonius Theorem for parallelogram: In a parallelogram, the sum of squares ofdiagonals is equal to twice the sum of the squares of its adjacent sides i.e. if a & b are lengths of two adjacent sides AB & BC respectively and d_1 & d_2 are the lengths of diagonals AC and BD respectively in a parallelogram *ABCD* (as shown in figure) then,

$$d_1^2 + d_2^2 = 2\left(a^2 + b^2\right)$$

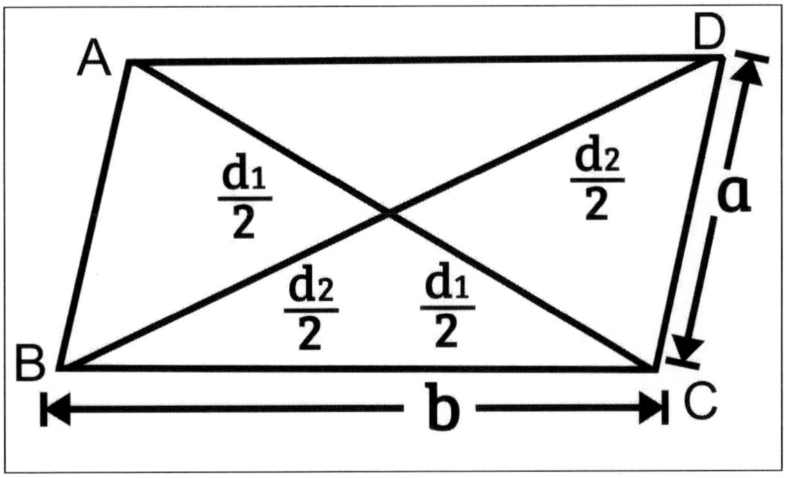

Parallelogram ABCD with adjacent sides a, b, & diagonals d_1, d_2.

Proof: Consider a parallelogram ABCD having adjacent sides AB = a & BC = b and the diagonals AC = d_1 & D= d_2. Let the diagonals AC and BD be bisecting each other at the point O (as shown in the fig-6) then we have,

$$AO = OC = \frac{d_1}{2} \ \& \ BO = OD = \frac{d_2}{2}$$

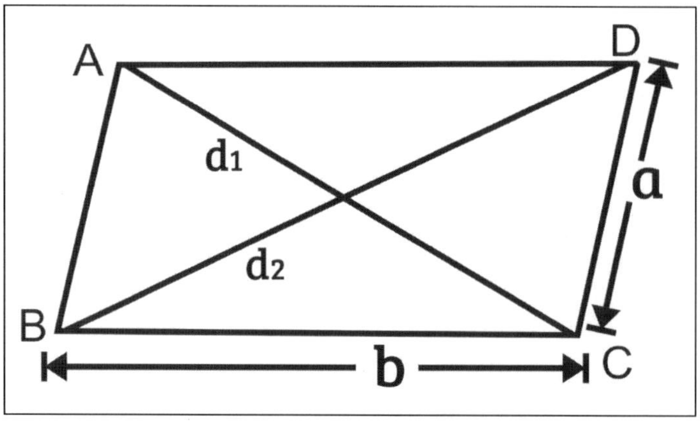

In parallelogram $ABCD$, $AO = OC = d_1/2$ & $BO = OD\ d_2/2$.

In ΔABC, BO is median. Now, applying Apollonius theorem in this ΔABC as follows,

$$AB^2 + BC^2 = \frac{AC^2}{2} + 2(BO)^2$$

$$a^2 + b^2 = \frac{d_1^2}{2} + 2\left(\frac{d_2}{2}\right)^2$$

$$a^2 + b^2 = \frac{d_1^2}{2} + \frac{d_2^2}{2}$$

$$a^2 + b^2 = \frac{d_1^2 + d_2^2}{2}$$

$$d_1^2 + d_2^2 = 2\left(a^2 + b^2\right) \qquad \text{Proved.}$$

Special Case 1: A rhombus has all four sides equal hence setting $b = a$ in above formula of parallelogram, we get,

$$d_1^2 + d_2^2 = 2\left(a^2 + a^2\right)$$

$$d_1^2 + d_2^2 = 4a^2$$

The above result can also be obtained by using Pythagoras theorem in a rhombus.

Special Case 2: A rectangle has both diagonals equal hence setting $d_1 = d_2 = d$ in above formula of parallelogram, we get,

$$d^2 + d^2 = 2\left(a^2 + b^2\right)$$

$$2d^2 = 2\left(a^2 + b^2\right)$$

$$d^2 = a^2 + b^2$$

The above result is true for a rectangle by Pythagoras theorem.

British Flag Theorem

The British flag theorem says that if a point P is chosen inside rectangle ABCD then $AP^2 + CP^2 = BP^2 + DP^2$. The theorem is called the British flag theorem due to the similarities between the British flag and a diagram of the points (shown below):

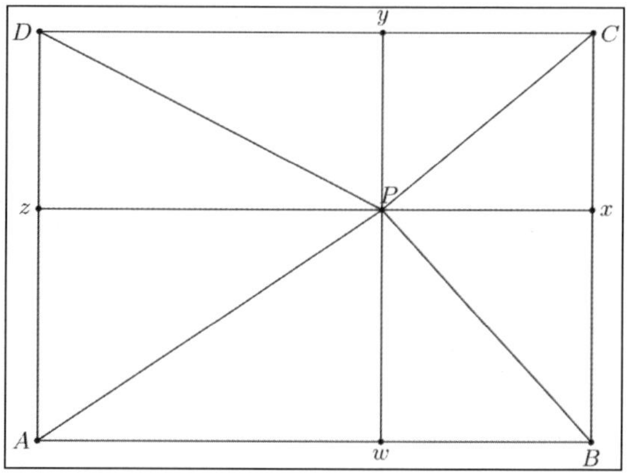

The theorem also applies if the point *P* is selected outside or on the boundary of the rectangle, although the proof is harder to visualize in this case.

Proof

Squares of lengths suggest right triangles. We build right triangles by drawing a line through P perpendicular to two sides of the rectangle, as shown below. Both *AXYD* and *BXYC* are rectangles.

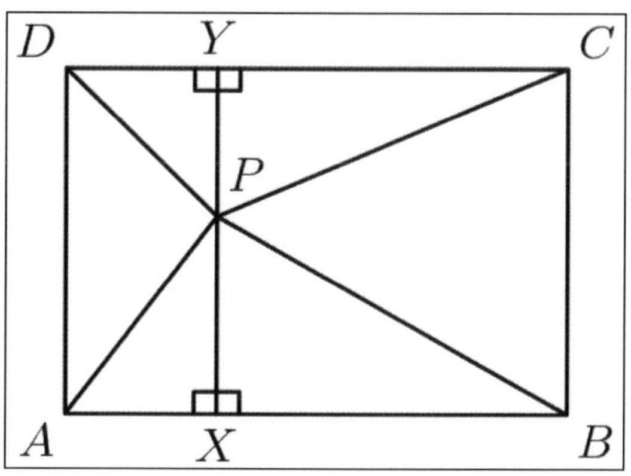

Applying the Pythagorean Theorem to each of the four right triangles in the diagram we have,

$$PA^2 = AX^2 + XP^2,$$

$$PB^2 = BX^2 + XP^2,$$

$$PC^2 = CY^2 + YP^2,$$

$$PD^2 = DY^2 + YP^2.$$

So, we have:

$$PA^2 + PC^2 = AX^2 + XP^2 + CY^2 + YP^2,$$
$$PB^2 + PD^2 = BX^2 + XP^2 + DY^2 + YP^2.$$

From rectangles *AXYD* and *BXYC*, we have *AX* = *DY* and *BX* = *CY*, so the expressions above for $PA^2 + PC^2$ and $PB^2 + PD^2$ are equal, as desired.

References

- Euclidean-geometry: cs.mcgill.ca, Retrieved 08 July, 2019
- Euclids-postulates, maths-introduction-to-euclids-geometry: toppr.com, Retrieved 10 June, 2019
- Euclidean-plane: theochem.ru.nl, Retrieved 04 April, 2019
- Curvature-plane-curves: math24.net, Retrieved 05 March, 2019
- Parallelogram-law, maths: byjus.com, Retrieved 09 February, 2019
- Proofs-of-Apollonius-Theorem-2-D-Geometry-by-HCR- 38953628: academia.edu, Retrieved 09 June, 2019
- British-Flag-Theorem: artofproblemsolving.com, Retrieved 16 August, 2019

3

Non-Euclidean Geometry: A Comprehensive Study

Non-Euclidean geometry is based on principles that are closely related to the Euclidean geometry. It includes elliptic geometry, spherical geometry, hyperbolic geometry. Concepts of hypercycles and horocycles, Apeirogon, tessellations, Gaussian and mean curvatures, etc. are also studied within it. This chapter has been carefully written to provide an easy understanding of non-Euclidean geometry.

Non-Euclidean geometry refers to certain types of geometry which differ from plane and solid geometry which dominated the realm of mathematics for several centuries. There are other types of geometry which do not assume all of Euclid's postulates such as hyperbolic geometry, elliptic geometry, spherical geometry, descriptive geometry, differential geometry, geometric algebra, and multidimensional geometry. These geometries deal with more complex components of curves in space rather than the simple plane or solids used as the foundation for Euclid's geometry. The first five postulates of Euclidean geometry will be listed in order to better understand the changes that are made to make it non-Euclidean.

1. A straight line can be drawn from any point to any point.
2. A finite straight line can be produced continuously in a straight line.
3. A circle may be described with any point as center and any distance as a radius.
4. All right angles are equal to one another.
5. If a transversal falls on two lines in such a way that the interior angles on one side of the transversal are less than two right angles, then the lines meet on the side on which the angles are less than two right angles.

A consistent logical system for which one of these postulates is modified in an essential way is non-Euclidean geometry. Although there are different types of Non-Euclidean geometry which do not use all of the postulates or make alterations of one or more of the postulates of Euclidean geometry, hyperbolic and elliptic are usually most closely associated with the term non-Euclidean geometry.

Hyperbolic geometry is based on changing Euclid's parallel postulate, which is also referred to as Euclid's fifth postulate, the last of the five postulates of Euclidian Geometry. Euclid's parallel postulate may also be stated as one and only one parallel to a given line goes through a given point not on the line.

Elliptic geometry uses a modification of Postulate II. Postulate II allows for lines of infinite length, which are denied in Elliptic geometry, where only finite lines are assumed.

Elliptic Geometry

Elliptic Geometry is a type of non-Euclidean geometry, that is, a geometric theory based on axioms whose requirements are to a considerable degree different from the requirements of the axioms of Euclidean geometry. The basic objects, or elements, of three dimensional elliptic geometry are points, lines, and planes; the basic concepts of elliptic geometry are the concepts of incidence (a point is on a line, a line is in a plane), order (for example, the order of points on a line or the order of lines passing through a given point in a given plane), and congruence (of figures).

The requirements of the axioms of elliptic geometry concerning membership and order completely coincide with the requirements of the axioms of projective geometry. Accordingly, such propositions as the following are found in elliptic geometry: through any two points there passes one line; any two planes intersect in one line; any two coplanar lines intersect in one point; and the points on a line are arranged in a cyclic order, as are lines lying in the same plane and passing through the same point.

The requirements of the axioms of elliptic geometry concerning congruence are similar to the corresponding requirements of the axioms of Euclidean geometry; in any case, they provide for motions of figures in the elliptic plane and in elliptic space that are just as free as in the plane and space of Euclidean geometry.

The metric properties of the elliptic plane coincide "in the small" with the metric properties of the ordinary sphere. More precisely, for any point of the elliptic plane there exists a part of the plane, containing this point, that is isometric to some part of the sphere; the radius R of this sphere is the same for all planes in the given elliptic space. The number $K = 1/R^2$ is called the curvature of the elliptic space. The smaller is K, the closer are the properties of this space to those of Euclidean space.

The properties of the elliptic plane "in the large" differ from the properties of the sphere as a whole. For example, in the elliptic plane, two lines intersect in one point; on the sphere, two great circles, which play the role of lines in spherical geometry, intersect in two points. A line in a plane does not separate the planethat is, if the line a is in the plane α, then any two points of α not on a can be joined by a line segment that does not intersect a.

Elliptic Line

In Elliptic space E_1 the norm of $\boldsymbol{a} = d\boldsymbol{e}_0 + a\boldsymbol{e}_1$ is given by $\|\boldsymbol{a}\| = \sqrt{d^2 + a^2}$, so any non-zero vector can be normalised. A normalised vector can be written as,

$$\boldsymbol{a} = -\boldsymbol{e}_0 \sin \alpha + \boldsymbol{e}_1 \cos \alpha.$$

It dually represents a point in E_1 at $x = \tan \alpha$ if $\alpha \neq \frac{\pi}{2} + \pi k$ where $k \in \mathbb{Z}$, otherwise $\boldsymbol{a} = \pm\, e_0$. Note that $\boldsymbol{e}_1\boldsymbol{a} = \cos\alpha + \boldsymbol{e}_{01} \sin \alpha = e^{\alpha e01}$ and, therefore, a can be expressed via the exponential function as follows:

$$\boldsymbol{a} = \boldsymbol{e}_1 e^{\alpha e_{01}}.$$

In the following, we identify vectors in the dual model space $\mathbb{R}^{2*}$ with the points in E_1 they dually represent. We will refer to **a**, **b**, etc as points in E_1.

The distance $r \in \left[0, \frac{\pi}{2}\right]$ between two normalised points **a** and **b** is defined by,

$$\sin\ r = |\boldsymbol{a} \vee \boldsymbol{b}|.$$

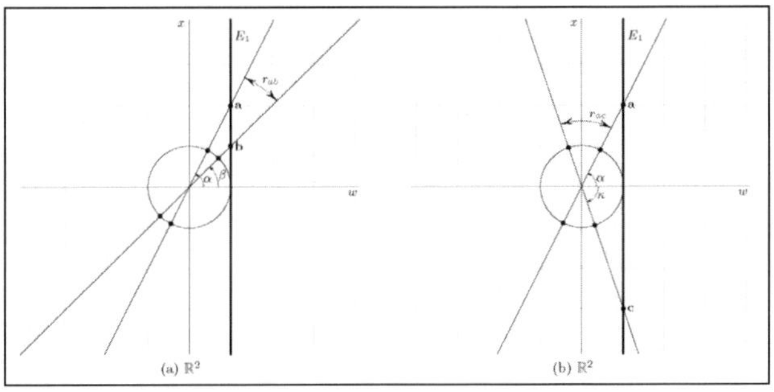

The distance between points in E_1.

In $E_1, |\boldsymbol{a} \cdot \boldsymbol{b}|^2 + |\boldsymbol{a} \vee \boldsymbol{b}|^2 = 1$ for the normalised points and, therefore, the distance also satisfies $\cos r = |\boldsymbol{a} \cdot \boldsymbol{b}|$. For the normalised points $\boldsymbol{a} = -\boldsymbol{e}_0 \sin\alpha + \boldsymbol{e}_1 \cos\alpha$ and $b = -\boldsymbol{e}_0 \sin\beta + \boldsymbol{e}_1 \cos\beta$, we get $\boldsymbol{a} \cdot \boldsymbol{b} = \sin\alpha \sin\beta + \cos\alpha \cos\beta = \cos(\alpha - \beta)$ and, therefore,

$$r = \begin{cases} |\alpha - \beta|, \ if\, |\alpha - \beta| \leq \frac{\pi}{2}, \\ \pi - |\alpha - \beta|, \ if\ \frac{\pi}{2} < |\alpha - \beta| \leq \pi. \end{cases}$$

The distance in E_1 has a maximum value of $\frac{\pi}{2}$. For instance, the distance from the origin $\boldsymbol{e}_1$ to $\boldsymbol{e}_0$ equals $\frac{\pi}{2}$, so $\boldsymbol{e}_0$ is not at infinity in E_1. In fact, $\mathbf{e}_0$ is not different from any other point in E_1.

The points $\boldsymbol{a} = \frac{1}{\sqrt{5}}(-2\boldsymbol{e}_0 + \boldsymbol{e}_1)$ and $\boldsymbol{b} = \frac{1}{\sqrt{2}}(-\boldsymbol{e}_0 + \boldsymbol{e}_1)$ shown in figure are parametrised by α = arctan 2, β = $\frac{\pi}{4}$ via, which gives $r_{ab} = \arctan 2 - \frac{\pi}{4}$ for the distance between them. Similarly, for the points $\boldsymbol{a} = \frac{1}{\sqrt{5}}(-2\boldsymbol{e}_0 + \boldsymbol{e}_1)$ and $\boldsymbol{c} = \frac{1}{\sqrt{10}}(3\boldsymbol{e}_0 + \boldsymbol{e}_1)$ shown in figure, we get $r_{ac} = \pi - (\arctan 2 + \arctan 3)$ for the distance between them (**c** is parametrised by $\kappa = -\arctan 3$). In both cases, the distance between the points is equal to the angle (in Euclidean sense in $\mathbb{R}^2$) between the linear subspaces representing the points. Due to this property of the distance measure in Elliptic space, it is convenient to visualise E_1 as a unit circle centered on the origin as shown in figure, instead of a straight line at $w = 1$. The circle is defined by,

$$w^2 + x^2 = 1$$

and the correspondence between the linear subspaces of $\mathbb{R}^2$ and the points on the circle is determined by the intersection of the subspaces with the circle. A linear subspace intersects the circle in two points, which by construction are identified and treated as a single point in E_1. In this representation of E_1, it is almost equivalent to a 1-dimensional version of spherical geometry denoted by $\mathbb{S}^1$. The difference between E_1 and $\mathbb{S}^1$ is that the antipodal points on the circle, which are treated as separate points in $\mathbb{S}^1$, are identical in E_1.

The polar point $\mathbf{ae}_{01}$ of a point $\mathbf{a}$ is a point in E_1 which is at a distance of $\frac{\pi}{2}$ from $\mathbf{a}$. This can be readily seen from $\mathbf{ae}_{01} = \|a\| \mathbf{e}_1 e^{\alpha e01} e^{\frac{\pi}{2} e_{01}} = \|a\| \mathbf{e}_1 e^{\left(\alpha + \frac{\pi}{2}\right) e_{01}}$, where we used $e_{01} = e^{\frac{\pi}{2} e_{01}}$. For instance, e_0 is the polar point of $-\mathbf{e}_1$, and $\mathbf{e}_1$ is the polar point of $\mathbf{e}_0$.

Spinors in E_1 are defined in the standard way as the product of an even number of normalised proper points. Note that every point in E_1 is proper, including $\mathbf{e}_0$. Any spinor can be written as $S = e^{\alpha e01}$ for some $\alpha \in \mathbb{R}$, e.g. $(-e_1) e_1 = -1 = e^{\pi e01}$. Spinors in E_1 can be identified with normalised complex numbers.

The translation of a point $\mathbf{a}$ by λ is given by $\mathrm{T}\mathbf{a}\mathrm{T}^{-1}$, where $T = e^{-\frac{1}{2} \lambda e_{01}}$. If $\mathbf{a}$ is parametrised with α, then $\mathbf{a}' = \mathrm{T}\mathbf{a}\mathrm{T}^{-1}$ can be parametrised with $\alpha' = \alpha + \lambda$, which follows from $\mathbf{a}' = \mathrm{T}\,\mathbf{a}\,\mathrm{T}^{-1} = \mathbf{a}\mathrm{T}^{-2} = \mathbf{e}_1 e^{\alpha e_{01}} e^{\lambda e_{01}} = \mathbf{e}_1 e^{(\alpha+\lambda) e_{01}}$ where we used $\mathbf{a} = \mathbf{e}_1 e^{\alpha e_{01}}$. Note that $\mathbf{a} = \mathbf{e}_1 e^{\alpha e_{01}}$ can be interpreted as the translation of $\mathbf{e}_1$, the origin of E_1, by α. So, a translation in E_1 is equivalent to a Euclidean rotation around the origin of $\mathbb{R}^2$. The space E_1 is closed and periodic with the period equal to π; translating any point by π yields the original point (albeit with the opposite orientation).

The top-down reflection of $\mathbf{a}$ in $\mathbf{b}$ is given by $-\mathbf{bab}^{-1}$. If a is parametrised with α and b with β, then $\mathbf{a}' = -\mathbf{bab}^{-1}$ can be parametrised with $\alpha' = \beta - (\alpha - \beta)$, which follows from $\mathbf{a}' = -\mathbf{bab}^{-1} = -e_1 e^{\beta e_{01}} \mathbf{e}_1 e^{\alpha e_{01}} e^{-\beta e_{01}} \mathbf{e}_1^{-1} = -e_1 e^{\beta e_{01}} e^{-\alpha e_{01}} e^{\beta e_{01}} = -e_1 e^{(\beta - \alpha + \beta) e_{01}}$. ± Note that the orientation of $\mathbf{a}$' is the opposite of $\mathbf{a}$.

The projection of $\mathbf{a}$ on $\mathbf{b}$ is given by the familiar expression $(\mathbf{a} \cdot \mathbf{b})\mathbf{b}^{-1}$, which yields $\mathbf{b}$ cos(α – β) if $\mathbf{a}$ and $\mathbf{b}$ are normalised ($\mathbf{a}$ is parametrised by α, and $\mathbf{b}$ by β). Indeed, $(\mathbf{a} \cdot \mathbf{b})\mathbf{b}^{-1} = \frac{1}{2}(\mathbf{ab} + \mathbf{ba})\mathbf{b} = \frac{1}{2}\mathbf{b}(\mathbf{ba} + \mathbf{ab}) = \boldsymbol{b}\frac{1}{2}\left(e^{(\alpha-\beta)e_{01}} + e^{(\beta-\alpha)e_{01}}\right) = \mathbf{b}\cos(\alpha - \beta)$. So, if $\mathbf{a}$ is at a distance of $\frac{\pi}{2}$ from $\mathbf{b}$, the projection of $\mathbf{a}$ on $\mathbf{b}$ is zero. Unlike $\mathbf{e}_0$ in Euclidean space, $\mathbf{e}_0$ in E_1 is invertible and it is possible to project on $\mathbf{e}_0$ as on any other point. The rejection of $\mathbf{a}$ by $\mathbf{b}$ is given by $(\mathbf{a} \wedge \mathbf{b})\mathbf{b}^{-1}$, which for the normalised points yields $\mathbf{be}_{01}$ sin(α – β). So, the rejection by $\mathbf{b}$ coincides with the polar point of $\mathbf{b}$ weighted by sin(α – β). Moreover, the rejection is equivalent to projection on the polar point. The expression for scaling, which we used in degenerate spaces, is not applicable in a non-degenerate space.

Elliptic Plane

The norm of a line $\mathbf{a} = d\mathbf{e}_0 + a\mathbf{e}_1 + b\mathbf{e}_2$ and a point $\boldsymbol{P} = w\boldsymbol{e}_{12} + x\boldsymbol{e}_{20} + y\boldsymbol{e}_{01}$ is given by $\|\boldsymbol{a}\| = \sqrt{d^2 + a^2 + b^2}$ and $\|\boldsymbol{P}\| = \sqrt{w^2 + x^2 + y^2}$. So, any line or point in E_2 can be normalised. The distance $r \in \left[0, \frac{\pi}{2}\right]$ between two normalised points $\mathbf{P}$ and $\mathbf{Q}$ is defined by,

$$\sin r = \|\boldsymbol{P} \vee \boldsymbol{Q}\|.$$

Since $|\boldsymbol{P}\cdot\boldsymbol{Q}|^2+\|\boldsymbol{P}\vee\boldsymbol{Q}\|^2=1$ for normalised points in E_2, the distance r also satisfies $\cos r=|\boldsymbol{P}\cdot\boldsymbol{Q}|$. If the point **P** is expressed via the standard coordinates x and y, i.e. $\boldsymbol{P}=\boldsymbol{e}_{12}+x\boldsymbol{e}_{20}+y\boldsymbol{e}_{01}$, then $\|\boldsymbol{P}\|=\sqrt{1+x^2+y^2}$ and,

$$\sin r=\frac{\sqrt{x^2+y^2}}{\sqrt{1+x^2+y^2}},\quad \cos r\frac{1}{\sqrt{1+x^2+y^2}}.$$

So, those points that are at an infinite distance from the origin in Euclidean space $\mathbb{E}^2$ are at the distance of $\frac{\pi}{2}$ from the origin in Elliptic space E_2, which is the maximum distance in E_2. The distance between arbitrary points P and Q can be understood as the Euclidean angle between the 1-dimensional linear subspaces in $\mathbb{R}^3$ corresponding to J(**P**) and J(**Q**), where $\mathbb{R}^3$ is viewed as a 3-dimensional Euclidean space, where $\boldsymbol{P}=\boldsymbol{e}_{12}+\boldsymbol{e}_{20}$ and $\boldsymbol{Q}=\boldsymbol{e}_{12}+2\boldsymbol{e}_{01}$, which upon the normalisation yields $r=\arccos\left(\frac{1}{\sqrt{2}}\frac{1}{\sqrt{5}}\right)$ for the distance between the points.

The angle $\alpha\in[0,\pi]$ between two normalised lines a and b, or more precisely the angle between the orientation vectors of the lines, is defined by:

$$\cos\alpha=\boldsymbol{a}\cdot\boldsymbol{b}$$

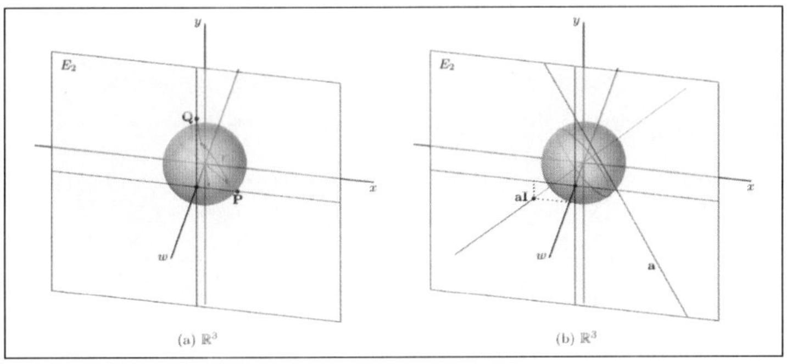

Visualising E_2.

It corresponds to the Euclidean angle between **a** and **b** only if both lines pass through the origin. In general, α is equal to the Euclidean angle between the 2-dimensional linear subspaces in $\mathbb{R}^3$ corresponding to *J*(**a**) and *J*(**b**), or more precisely it is equal to the Euclidean angle between the vectors *I*(**a**) and *I*(**b**), which are perpendicular (in the Euclidean sense of $\mathbb{R}^3$) to the planes *J*(**a**) and *J*(**b**). Since both the distance and angular measures are elliptic, there is a close relationship between them. For instance, if $0\le\alpha\le\frac{\pi}{2}$, then α is equal to the distance between the polar points of the lines a and b, i.e. cos $\alpha=|(\mathbf{aI})\cdot(\mathbf{bI})|$.

Due to the above mentioned properties of the distance and angular measures, it might be more convenient to visualise E_2 as a unit sphere centered on the origin in $\mathbb{R}^3$ rather than a plane at w = 1. The sphere is defined by,

$$w^2+x^2+y^2=1$$

and its antipodal points are identified. This is a natural extension of the representation of E_1 as a unit circle.

When E_2 is viewed as the unit sphere, each line in E_2 can be identified with a great circle of the sphere. For instance, the line $\boldsymbol{a} = \boldsymbol{P} \vee \boldsymbol{Q} = -2\boldsymbol{e}_0 + 2\boldsymbol{e}_1 + \boldsymbol{e}_2$ shown in figure is represented by a 2-dimensional linear subspace of $\mathbb{R}^3$, which intersects the unit sphere along a great circle. If the great circle is thought of as the equator, then the polar point $\boldsymbol{aI} = -2\boldsymbol{e}_{12} + 2\boldsymbol{e}_{20} + \boldsymbol{e}_{01} = -2\left(\boldsymbol{e}_{12} - \boldsymbol{e}_{20} - \frac{1}{2}\boldsymbol{e}_{01}\right)$ of the line **a** is located at the poles of the sphere with respect to the equator. The polar point **aI** is at the distance of $\frac{\pi}{2}$ from any point on the line **a**. Hence, the commutator **P** × **Q**, which coincides with the polar point of the line **P** ∨ **Q**, is located at a finite distance from the origin. Since $\mathbf{I}^{-1} = -\mathbf{I}$ in E_2, the definition of the duality transformation implies that $J(\mathbf{a}) = -I(\mathbf{aI})$, where J and I are the duality and identity transformations. So, a line a and its polar point aI are related to each other in E_2 in essentially the same way as a line and a point dual to each other in the sense of projective duality.

The distance r between a normalised line **a** and a normalised point **P** satisfies $\sin r = |\boldsymbol{a} \vee \boldsymbol{P}|$ and alternatively $\cos r = \|\boldsymbol{a} \cdot \boldsymbol{P}\|$. For instance, the distance from $\boldsymbol{P} = \boldsymbol{e}_{12} - \frac{3}{5}\boldsymbol{e}_{20} + \frac{4}{5}\boldsymbol{e}_{01}$ to $\boldsymbol{a} = -2\boldsymbol{e}_0 + 2\boldsymbol{e}_1 + \boldsymbol{e}_2$ is given by $\sin r = \frac{2.4}{3\sqrt{2}}$, where normalisation has been taken into account. For a normalised line $\boldsymbol{a} = d\boldsymbol{e}_0 + a\boldsymbol{e}_1 + b\boldsymbol{e}_2$, we get $\sin r = |d|$ for the distance r from the line to the origin.

For instance, the distance of $\mathbf{e}_0$ from the origin equals $\frac{\pi}{2}$.

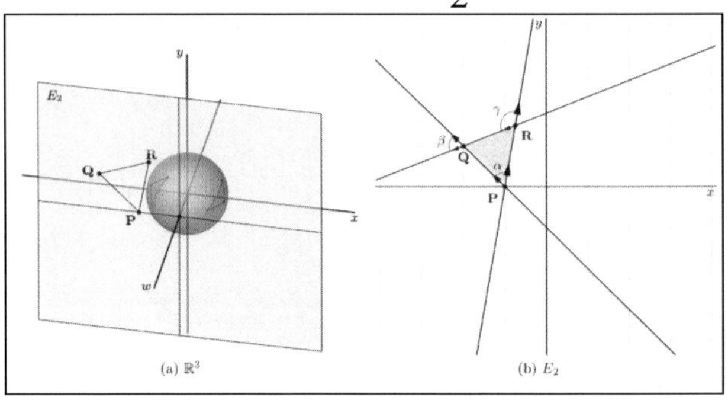

Triangles in E_2 (1).

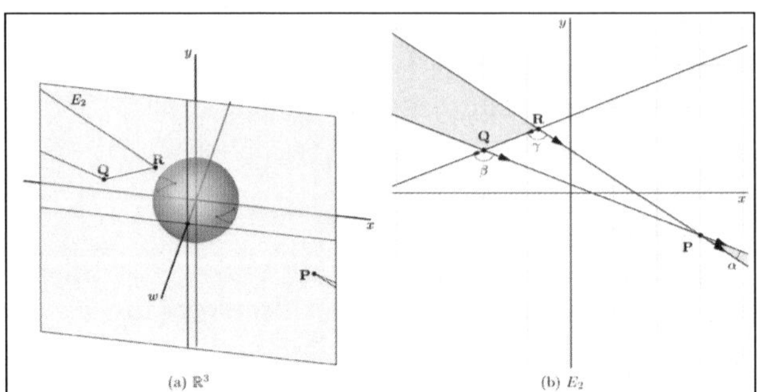

Triangles in E_2 (2).

Unlike the flat spaces $\mathbb{E}^2$ and $\mathbb{M}^2$, there is no one-to-one correspondence between the area S of a triangle in E_2 defined by three normalised points **P**, **Q**, **R** and the value of $|P \vee Q \vee R|$. However, if one of the angles of the triangle is right, say, the angle at the point P equals $\frac{\pi}{2}$, then the area S satisfies,

$$\sin S = \frac{|P \vee Q \vee R|}{1 + |Q \cdot R|}$$

where **P**, **Q**, and **R** are assumed to be normalised and $(P \vee Q) \cdot (P \vee R) = 0$. In particular, if all three points are at the distance $\frac{\pi}{2}$ from one another, then sin S = 1 and $S = \frac{\pi}{2}$, which is the maximum possible area of a triangle in E_2. The total area of E_2 equals 2π.

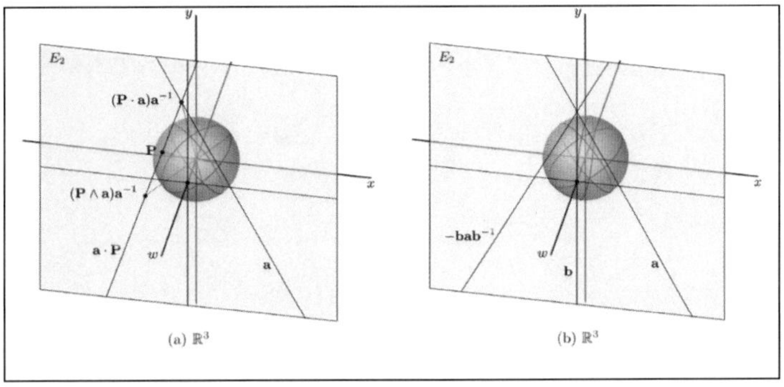

Some basic properties of E_2.

An example of a triangle is shown in figure, where $P = \frac{1}{\sqrt{2}}(e_{12} - e_{20}), Q = \frac{1}{\sqrt{6}}(e_{12} - 2e_{20} + e_{01})$, and $R + \frac{4}{\sqrt{61}}\left(e_{12} - \frac{3}{4}e_{20} + \frac{3}{2}e_{01}\right)$. The angles associated with the triangle can be found by applying the definition, which gives $\alpha = \arccos\left(\frac{r \cdot q}{\|r\|\|q\|}\right), \beta = \arccos\left(\frac{r \cdot p}{\|r\|\|p\|}\right), \gamma = \arccos\left(\frac{q \cdot p}{\|q\|\|p\|}\right)$ where $r = P \vee Q,\ q = P \vee R,\ p = R \vee Q$. Observe the correspondence between the angles as shown in figure and the bottom-up orientation of lines, which constitute the sides of the triangle. Then, the area S can be computed with,

$$S = \alpha + \beta - \gamma,$$

which is equivalent to the spherical excess formula $S = \alpha + \beta + (\pi - \gamma) - \pi$. The same area can be computed by breaking the triangle into two right triangles and computing the area of each by applying $\sin S = \frac{|P \vee Q \vee R|}{1 + |Q \cdot R|}$.

Another example is shown in figure, where **Q** and **R** are the same and $P = \frac{1}{\sqrt{11}}(e_{12} + 3e_{20}e_{01})$. Note the unusual appearance of this triangle in the plane E_2. The lines by the sides of the triangle are defined by $r = Q \vee P$, $q = R \vee P$, $p = R \vee Q$, which gives the bottom-up orientation shown in figure. Then, $S = \alpha + \beta - \gamma$ as before.

The line passing through **P** and perpendicular to **a** is given by the inner product $a \cdot P$ as usual, where $P = e_{12} - \frac{3}{5}e_{20} + \frac{4}{5}e_{01}$ and $a = -2e_0 + 2e_1 + e_2$. Note that $\mathbf{a} \cdot \mathbf{P} = 0$ if the location of P coincides with that of the polar point **aI** of **a**. Projection, rejection, and reflection are defined in the usual way and since any non-zero blade is invertible in E_2 it is possible to project on the line $\mathbf{e}_0$ and, say, the point $\mathbf{e}_{20}$, for example. For an illustration of the projection and rejection of a point by a line. Note that the rejection $(P \wedge a)a^{-1}$ is at the distance of $\frac{\pi}{2}$ from **a**, so it is located at the same position in E_2 as the polar point of **a**. The top-down reflection of a line **a** in a line **b** is given by $-\mathbf{bab}^{-1}$ and is illustrated in figure, where $a = 2e_0 + 2e_1 + e_2$ and $b = e_1$. In general, in E_2 the projection of B_l on A_k is given by $(B_l \cdot A_k)A_k^{-1}$ and the rejection of B_l by A_k is given by $(B_l \wedge A_k)A_k^{-1}$. The top-down reflection of B_l in A_k is given by $(-1)^{kl} A_k B_l A_k^{-1}$ and the bottom-up reflection by $A_k B_l A_k^{-1}$.

Note that $a \cdot P$ passes through the polar point **aI** of the line **a**. In fact, any line in E_2 passing through **aI** is perpendicular to **a**. Moreover, the polar point of any line perpendicular to **a** lies on the line **a**. This can be used to construct a triangle with maximal area of $\frac{\pi}{2}$ as follows. For any line **a**, select a line **b** passing through **aI**, then define the third line $c = (aI) \vee (bI)$. The angles of the resulting triangle are all equal to $\frac{\pi}{2}$ and, therefore, its area equals $\frac{\pi}{2}$. Observe also that in elliptic space E_2 any two lines intersect at a single point, which is at a finite distance from the origin. Hence, there are no parallel lines in E_2.

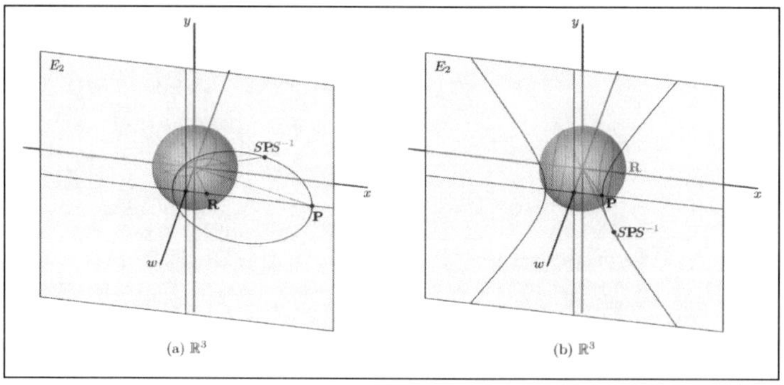

Rotation in E_2

The spin group of E_2 is defined in the standard way and consists of spinors of the form e^A where A is a bivector. It is isomorphic to the group of normalised quaternions. The action $S A_k S^{-1}$ of a spinor $S = e^{-\frac{1}{2}\alpha R}$ on the geometric object dually represented by the blade A_k is consistent with the rotation of A_k around the point **R** by the angle α, where α refers to the angle according to the elliptic metric, rather than Euclidean. No translation is possible in E_2 since there are no points at infinity. The rotation of the point $P = e_{12} + 3e_{20}$ around $R = e_{12} + \frac{1}{2}e_{20}$ by the angle $\alpha = \frac{\pi}{4}$ is shown in figure: The solid curve in E_2 shows the trajectory of **P** under the action of the spinor $S = e^{-\frac{1}{2}tR}$, where the parameter t varies from 0 to 2π (the projection of this curve onto the unit sphere is shown with the solid curve as well). The curve consists of points at an equal distance from R and, therefore, forms a circle in E_2, which passes through **P** and whose centre is at **R**. The appearance of the circle on the plane E_2 is elliptic.

Figure shows another example of the rotation, where $\boldsymbol{P} = \boldsymbol{e}_{12} + \frac{2}{3}\boldsymbol{e}_{20}$, $\mathbf{R} = \boldsymbol{e}_{20}$, and $\alpha = \frac{\pi}{4}$. Note that R is shown on the unit sphere only; it corresponds to a stack of lines in the plane E_2. The solid curve shows the trajectory of **P** under the action of the spinor $S = e^{-\frac{1}{2}tR}$, where the parameter t varies from 0 to 2π. It is also a circle but its appearance in the plane E_2 is hyperbolic.

In general, a circle in E_2 could appear as an ellipse, two branches of a hyperbola, a parabola, or a straight line. The latter represents a circle whose radius equals $\frac{\pi}{2}$. The circle is parabolic if there is only one point on it that belongs to the line $\mathbf{e}_0$ and hyperbolic if there are two points on the circle that belong to $\mathbf{e}_0$. Otherwise, it is elliptic. The circle of radius $\frac{\pi}{2}$ whose centre is at the origin cannot be visualised on the plane E_2; this circle corresponds to the line $\mathbf{e}_0$.

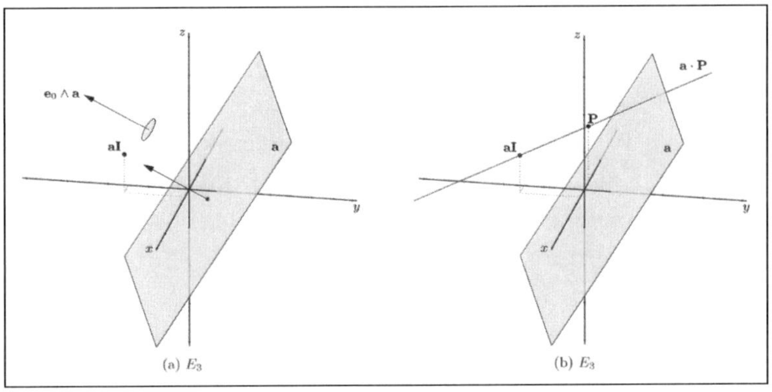

Basic properties of points and planes in E_3.

Elliptic Space

For a plane $\boldsymbol{a} = d\boldsymbol{e}_0 + a\boldsymbol{e}_1 + b\boldsymbol{e}_2 + c\boldsymbol{e}_3$, a line $\boldsymbol{\Lambda} = p_{10}\boldsymbol{e}_{10} + p_{20}\boldsymbol{e}_{20} + p_{30}\boldsymbol{e}_{30} + p_{23}\boldsymbol{e}_{23} + p_{12}\boldsymbol{e}_{12}$, where $p_{10}p_{23} + p_{20}p_{31} + p_{30}p_{12} = 0$, and a point $\mathbf{P} = w\boldsymbol{e}_{123} + x\boldsymbol{e}_{320} + y\boldsymbol{e}_{130} + z\boldsymbol{e}_{210}$, the norm is given by $\|\mathbf{a}\| = \sqrt{d^2 + a^2 + b^2 + c^2}$, $\|\boldsymbol{\Lambda}\| = \sqrt{p_{10}^2 + p_{20}^2 + p_{30}^2 + p_{23}^2 + p_{31}^2 + p_{12}^2}$, $\|\mathbf{P}\| = \sqrt{w^2 + x^2 + y^2 + z^2}$, and $\mathbf{a}^2 = \|\mathbf{a}\|^2$, $\boldsymbol{\Lambda}^2 = -\|\boldsymbol{\Lambda}\|^2$, $\mathbf{P}^2 = -\|\mathbf{P}\|^2$. Note that the above formula for $\|\boldsymbol{\Lambda}\|$ is only valid for lines; it is not applicable to non-simple bivectors.

The hyperplane E_3 is embedded in the model space $\mathbb{R}^4$ at w = 1. By analogy with the 1- and 2-dimensional cased, the 3-dimensional elliptic space can also be thought of as the unit 3-sphere centred on the origin of $\mathbb{R}$, which is defined by,

$$w^2 + x^2 + y^2 + z^2 = 1.$$

The antipodal points in the unit 3-sphere are identified. The immediate 3-dimensional vicinity of the origin of E_3 is near the unit 3-sphere and they resemble each other closely. As one moves away from the origin of E_3, the unit 3-sphere deviates from the hyperplane E_3, curving into the fourth dimension along w and towards the hyperplane w = 0. It is not possible to visualise the unit 3-sphere. However, one can infer the geometry in the 3-sphere by examining the geometry in E_3, which can be readily visualised.

Planes and lines in E_3 correspond to great spheres and great circles in the unit 3-sphere. For instance, the plane e_0 corresponds to the great sphere where the unit 3-sphere intersects the hyperplane w = 0. Any two distinct great spheres in the unit 3-sphere intersect along a great circle, which corresponds to the line at the intersect of two planes in E_3 corresponding to the great spheres. A point in E3 corresponds to a pair of antipodal points in the unit 3-sphere.

The polar point **aI** of a plane a is shown in figure, where $\mathbf{a} = e_0 + \frac{1}{2}e_1 - \frac{3}{2}e_2 + e_3$. The line $e_0 \wedge a$ is at the intersection of the planes **a** and $\mathbf{e}_0$. It is at a finite distance from the origin but it cannot be shown directly since it lies in $\mathbf{e}_0$ (it is depicted in the same way as a line at infinity in $\mathbb{E}^3$). A line passing through **P** and perpendicular to **a** is given by **a** · **P**, unless P is located at **aI** in which case **a** · **P** = 0. An example is shown in figure where $\mathbf{P} = \mathbf{e}_{123} - \frac{1}{2}\mathbf{e}_{320} + \frac{3}{2}\mathbf{e}_{210}$. Note that **a** · **P** passes through the polar point **aI** of a. In fact, any line perpendicular to a passes through **aI** and and conversely any line passing through **aI** is perpendicular to the plane **a**. The same observation applies to planes, i.e. **a** plane is perpendicular to **a** if and only if it passes through **aI**. This also concerns the plane passing through **aI** which would be parallel to a in Euclidean space (in elliptic space, there are no parallel planes since any two planes intersect at a finite distance from the origin). Note also that any line Λ that lies in the plane **a** is in the same relationship with respect to **aI** as an equatorial line and its polar point in E_2.

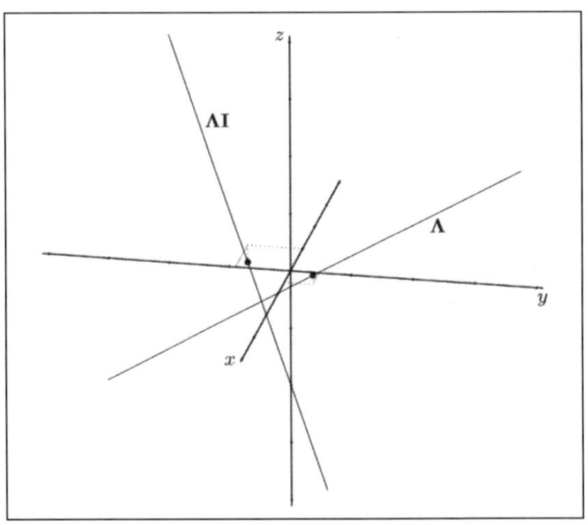

Lines in E_3

The distance $r \in \left[0, \frac{\pi}{2}\right]$ between normalised points **P** and **Q** is defined by,

$$\sin r \|\mathbf{P} \vee \mathbf{Q}\|$$

Where $\mathbf{P} \vee \mathbf{Q}$ is a line passing through **P** and **Q**. Since $\|\mathbf{P} \vee \mathbf{Q}\|^2 + |\mathbf{P} \cdot \mathbf{Q}|^2 = 1$ in $E_{3,}$ one also gets $\cos r = |\mathbf{P} \cdot \mathbf{Q}|$. If $\mathbf{P} = \mathbf{e}_{123} + x\mathbf{e}_{320} + y\mathbf{e}_{130} + z\mathbf{e}_{130} + z\mathbf{e}_{210}$ and $Q = \mathbf{e}_{123}$, i.e. **Q** is at the origin of E_3, then $\|\mathbf{P}\| = \sqrt{1 + x^2 + y^2 + z^2}$ and the distance from the origin to **P** satisfies,

$$\sin r \frac{\sqrt{x^2 + y^2 + z^2}}{\sqrt{1 + x^2 + y^2 + z^2}}, \quad \cos r \frac{1}{\sqrt{1 + x^2 + y^2 + z^2}}.$$

As in 1- and 2-dimensional cases, the maximum distance in E_3 equals $\frac{\pi}{2}$.

A plane **a** and its polar point **aI** are related by projective duality and J(a) = I(**aI**) where J and I are duality and identity transformations (note that $\mathbf{I}^{-1} = \mathbf{I}$ and $\mathbf{I}^2 = 1$ in E_3). Furthermore, the distance between **aI** and any point on **a** equals $\frac{\pi}{2}$. The polar point of the plane **PI** coincides with **P** and $J(\mathbf{P}) = I(\mathbf{PI})$. The distance from **P** to any point on the plane **PI** equals $\frac{\pi}{2}$. For a line $\mathbf{\Lambda}$, the pair of lines $\mathbf{\Lambda}$ and $\mathbf{\Lambda I}$ are also related by projective duality and ($J(\mathbf{\Lambda}) = I(\mathbf{\Lambda I})$). An example is shown in figure where $\mathbf{\Lambda} = -\frac{1}{3}\mathbf{e}_{20} + \mathbf{e}_{30} + \mathbf{e}_{23} - \mathbf{e}_{31} - \frac{1}{3}\mathbf{e}_{12}$. The distance between any point on $\mathbf{\Lambda}$ and any point on $\mathbf{\Lambda I}$ equals $\frac{\pi}{2}$.

The angle α between normalised planes **a** and **b**, or more precisely the angle between the top-down orientation vectors of the planes, is defined by,

$$\cos\alpha = \mathbf{a} \cdot \mathbf{b}.$$

For any normalised **a** and **b**, their geometric product yields **ab** = cos α + $\mathbf{\Lambda}$ sin α, where $\mathbf{\Lambda}$ is a normalised line at the intersection of the planes **a** and **b**.

A plane in E_3 can be considered as a copy of E_2, so the total area of any plane in E_3 is finite and is equal to 2π. The total volume of E_3 equals π². This is equivalent to half the volume of the unit 3-sphere in $\mathbb{R}^4$ with the Euclidean metric. A factor of $\frac{1}{2}$ is needed since antipodal points in the unit 3-sphere serving as a model of E_3 are identified. The volume of the unit 3-sphere in $\mathbb{R}^4$ should not be confused with the hypervolume of the 4-dimensional region bounded by the unit 3-sphere. There is no simple expression for the volume of an arbitrary 3-simplex, i.e. a tetrahedron, in E_3. However, the volume of a 3-simplex whose sides are perpendicular to one another equals $\frac{\pi^2}{8}$, which is the largest possible volume a 3-simplex can have in E_3.

The distance r from a normalised point **P** to a normalised plane **a** satisfies sin r = $|\mathbf{a} \vee \mathbf{P}|$ and cos r = $|\mathbf{a} \cdot \mathbf{P}|$. It equals the distance along the line $\mathbf{a} \cdot \mathbf{P}$ from **P** to the point where $\mathbf{a} \cdot \mathbf{P}$ intersects **a.** The distance from a normalised point **P** to a normalised line $\mathbf{\Lambda}$ satisfies sin r = $\|\mathbf{\Lambda} \vee \mathbf{P}\|$ and cos r = $\|\mathbf{\Lambda} . \mathbf{P}\|$, where $\mathbf{\Lambda} \vee \mathbf{P}$ is a plane passing through $\mathbf{\Lambda}$ and **P**, and $\mathbf{\Lambda} \cdot \mathbf{P}$ is a plane perpendicular to $\mathbf{\Lambda}$ and passing through **P**. Recall that $\mathbf{\Lambda} \vee \mathbf{P}$ if **P** lies on the line $\mathbf{\Lambda}$ and note that $\mathbf{\Lambda} \cdot \mathbf{P} = 0$ if **P** lies on $\mathbf{\Lambda I}$. The angle α between a line $\mathbf{\Lambda}$ and a plane **a** satisfies cos α = $\|\boldsymbol{a} . \mathbf{\Lambda}\|$ and sin α = $\|\mathbf{a} \wedge \mathbf{\Lambda}\|$, where $\mathbf{a}.\mathbf{\Lambda}$ is a plane passing through $\mathbf{\Lambda}$ and perpendicular to **a**, and $\mathbf{a} \wedge \mathbf{\Lambda}$ is a point where $\mathbf{\Lambda}$ intersects a (the angle α does not take into account the orientation of $\mathbf{\Lambda}$ and **a**).

Recall that $\mathbf{a} \wedge \boldsymbol{\Lambda} = 0$ if the line $\boldsymbol{\Lambda}$ lies in the plane $\mathbf{a}$. Note also that $\mathbf{a} \cdot \boldsymbol{\Lambda} = 0$ if the line $\boldsymbol{\Lambda}$ passes through $\mathbf{aI}$, in which case $\boldsymbol{\Lambda}\mathbf{I}$ lies in the plane $\mathbf{a}$. A plane is perpendicular to $\boldsymbol{\Lambda}$ if and only if it passes through $\boldsymbol{\Lambda}\mathbf{I}$. So, a set of planes perpendicular to $\boldsymbol{\Lambda}$ forms a sheaf of planes passing through $\boldsymbol{\Lambda}\mathbf{I}$. A line is perpendicular to a plane a if and only if it passes through $\mathbf{aI}$. The angle α between two normalised lines $\boldsymbol{\Lambda}$ and $\boldsymbol{\Theta}$ is given by $\boldsymbol{\Lambda} \cdot \boldsymbol{\Theta} = -\cos\alpha$ if the lines intersect.

The pair of lines $\boldsymbol{\Lambda}$ and $\boldsymbol{\Lambda}\mathbf{I}$ have an interesting property that the distance between $\boldsymbol{\Lambda}$ and $\boldsymbol{\Lambda}\mathbf{I}$ is constant everywhere along these lines. Such lines in E_3 are called Clifford-parallel. There are two families, positive and negative, of Clifford-parallel lines associated with the line $\boldsymbol{\Lambda}$. The lines in each family are Clifford-parallel to one another and $\boldsymbol{\Lambda}$ is Clifford-parallel to lines in both families. Each family foliates the whole elliptic space E_3. If one ignores weight and orientation, there are exactly two lines Clifford-parallel to $\boldsymbol{\Lambda}$ passing through every point in E_3, except for the points on the lines $\boldsymbol{\Lambda}$ and $\boldsymbol{\Lambda}\mathbf{I}$. One of these parallels is positive and the other is negative.

We assume the line $\boldsymbol{\Lambda}$ is normalised and is given by $\boldsymbol{\Lambda} = p_{10}\mathbf{e}_{10} + p_{20}\mathbf{e}_{20} + p_{30}\mathbf{e}_{30} + p_{23}\mathbf{e}_{23} + p_{31}\mathbf{e}_{31} + p_{12}\mathbf{e}_{12}$ with $p_{10}p_{23} + p_{20}p_{31} + p_{30}p_{12} = 0$. The positive family of lines Clifford-parallel to $\boldsymbol{\Lambda}$ is given by,

$$\boldsymbol{\Lambda}^+ = \boldsymbol{\Lambda}^+(\boldsymbol{\Omega}) = \boldsymbol{\Lambda} + (\boldsymbol{\Lambda}_m \cdot \boldsymbol{\Omega})(\mathbf{I} - 1)\boldsymbol{\Omega},$$

and the negative family by,

$$\boldsymbol{\Lambda}^- = \boldsymbol{\Lambda}^-(\boldsymbol{\Omega}) = \boldsymbol{\Lambda} + (\boldsymbol{\Lambda}_p \cdot \boldsymbol{\Omega})(\mathbf{I} + 1)\boldsymbol{\Omega},$$

Where,

$$\boldsymbol{\Lambda}_m = (p_{10} - p_{23})\mathbf{e}_{23} + (p_{20} - p_{31})\mathbf{e}_{31} + (p_{30} - p_{12})\mathbf{e}_{12},$$
$$\boldsymbol{\Lambda}_p = (p_{10} + p_{23})\mathbf{e}_{23} + (p_{20} + p_{31})\mathbf{e}_{31} + (p_{30} + p_{12})\mathbf{e}_{12},$$

and $\boldsymbol{\Omega}$ ranges over all normalised lines passing through the origin. The lines $\boldsymbol{\Lambda}_m$ and $\boldsymbol{\Lambda}_p$ also pass through the origin and are normalised since $\boldsymbol{\Lambda}$ is assumed to be normalised. The parallels $\boldsymbol{\Lambda}^+$, $\boldsymbol{\Lambda}^-$ are all normalised and their orientation is consistent with that of $\boldsymbol{\Lambda}$. It is convenient to parametrise $\boldsymbol{\Omega}$ with $\phi \in [0, 2\pi]$ and $\theta \in [0, \pi]$ by,

$$\boldsymbol{\Omega} = \boldsymbol{\Omega}_m(\phi,\theta) = e^{-\frac{1}{2}\phi\boldsymbol{\Lambda}_m} e^{-\frac{1}{2}\theta\boldsymbol{\Lambda}_m^{\perp}} \boldsymbol{\Lambda}_m e^{\frac{1}{2}\theta\boldsymbol{\Lambda}_m^{\perp}} e^{\frac{1}{2}\phi\boldsymbol{\Lambda}_m}$$

for the positive family and by,

$$\boldsymbol{\Omega} = \boldsymbol{\Omega}_p(\phi,\theta) = e^{-\frac{1}{2}\phi\boldsymbol{\Lambda}_p} e^{-\frac{1}{2}\theta\boldsymbol{\Lambda}_p^{\perp}} \boldsymbol{\Lambda}_p e^{\frac{1}{2}\theta\boldsymbol{\Lambda}_p^{\perp}} e^{\frac{1}{2}\phi\boldsymbol{\Lambda}_p}$$

for the negative family; $\boldsymbol{\Lambda}_m^{\perp}$ is perpendicular to $\boldsymbol{\Lambda}_m$, $\boldsymbol{\Lambda}_p^{\perp}$ is perpendicular to $\boldsymbol{\Lambda}_p$ and both $\boldsymbol{\Lambda}_m^{\perp}$ and $\boldsymbol{\Lambda}_p^{\perp}$ are normalised and pass through the origin. Then, $\left|\frac{\pi}{2} - \theta\right|$ gives the distance from $\boldsymbol{\Lambda}$ to the

Clifford parallels, both positive and negative, parametrised with (ϕ,θ). The exact meaning of ϕ depends on the choice of $\mathbf{\Lambda}_m^{\perp}$ and $\mathbf{\Lambda}_p^{\perp}$. Note that $\mathbf{\Lambda}_m \cdot \mathbf{\Omega}_m(\phi, \theta) = -\cos\theta$, i.e. θ in $\mathbf{\Omega} = \mathbf{\Omega}_m(\phi,\theta) = e^{-\frac{1}{2}\phi\Lambda_m} e^{-\frac{1}{2}\theta\Lambda_m^{\perp}} \mathbf{\Lambda}_m e^{\frac{1}{2}\theta\Lambda_m^{\perp}} e^{\frac{1}{2}\phi\Lambda_m}$ is the angle between $\mathbf{\Lambda}_m$ and $\mathbf{\Omega}_m(\phi, \theta)$, and $\mathbf{\Lambda}_p \cdot \mathbf{\Omega}_p(\phi,\theta) = -\cos\theta$, i.e. θ in $\mathbf{\Omega} = \mathbf{\Omega}_p(\phi,\theta) = e^{-\frac{1}{2}\phi\Lambda_p} e^{-\frac{1}{2}\theta\Lambda_p^{\perp}} \mathbf{\Lambda}_p e^{\frac{1}{2}\theta\Lambda_p^{\perp}} e^{\frac{1}{2}\phi\Lambda_p}$ is the angle between $\mathbf{\Lambda}_p$ and $\mathbf{\Omega}_p(\phi,\theta)$. So, the Clifford parallels can be given by,

$$\mathbf{\Lambda}^+ = \mathbf{\Lambda}^+(\phi,\theta) = \mathbf{\Lambda} - \cos\theta(\mathbf{I}-1)\mathbf{\Omega}_m(\phi,\theta),$$
$$\mathbf{\Lambda}^- = \mathbf{\Lambda}^-(\phi,\theta) = \mathbf{\Lambda} - \cos\theta(\mathbf{I}+1)\mathbf{\Omega}_p(\phi,\theta).$$

If $\mathbf{\Lambda}$ is not normalised, one can find the Clifford parallels of $\mathbf{\Lambda}/\|\mathbf{\Lambda}\|$ by applying the above formulas and then scale the parallels by $\|\mathbf{\Lambda}\|$ to get the Clifford parallels of $\mathbf{\Lambda}$. Some positive and negative Clifford parallels of $\mathbf{\Lambda} = \frac{3}{\sqrt{29}}\left(-\frac{1}{3}\mathbf{e}_{20} + \mathbf{e}_{30} + \mathbf{e}_{23} - \mathbf{e}_{31} - \frac{1}{3}\mathbf{e}_{12}\right)$ located at the distance $r = \frac{\pi}{10}$ from $\mathbf{\Lambda}$ are shown in figure (positive in (a) and negative in (b)).

Note that $\mathbf{\Lambda I}$ is a positive Clifford parallel of $\mathbf{\Lambda}$ (substitute $\mathbf{\Omega} = \mathbf{\Lambda}_m$ into $\mathbf{\Lambda}^+ = \mathbf{\Lambda}^+(\Omega) = \mathbf{\Lambda} + (\mathbf{\Lambda}_m \cdot \mathbf{\Omega})(\mathbf{I}-1)\mathbf{\Omega}$,) and $-\mathbf{\Lambda I}$ is a negative Clifford parallel of $\mathbf{\Lambda}$ (substitute $\mathbf{\Omega} = \mathbf{\Lambda}_p$ into $\mathbf{\Lambda}^- = \mathbf{\Lambda}^-(\Omega) = \mathbf{\Lambda} + (\mathbf{\Lambda}_p \cdot \mathbf{\Omega})(\mathbf{I}+1)\mathbf{\Omega}$,), while $\mathbf{\Lambda}$ belongs to both positive and negative families. Also, even though $\mathbf{\Lambda}_m$ is used in the definition of the positive Clifford parallels, $-\mathbf{\Lambda}_m$ is a negative Clifford parallel and $\mathbf{\Lambda}_p$ is a positive Clifford parallel. They can be obtained by substituting $\Omega = (\mathbf{\Lambda}_p + \mathbf{\Lambda}_m)/\|\mathbf{\Lambda}_p + \mathbf{\Lambda}_m\|$ into $\mathbf{\Lambda}^+ = \mathbf{\Lambda}^+(\Omega) = \mathbf{\Lambda} + (\mathbf{\Lambda}_m \cdot \mathbf{\Omega})(\mathbf{I}-1)\mathbf{\Omega}$, and $\mathbf{\Lambda}^- = \mathbf{\Lambda}^-(\Omega) = \mathbf{\Lambda} + (\mathbf{\Lambda}_p \cdot \mathbf{\Omega})(\mathbf{I}+1)\mathbf{\Omega}$.

Since $(\mathbf{I}+1)(\mathbf{I}-1) = 0$, the positive and negative Clifford parallels of $\mathbf{\Lambda}$ satisfy,

$$(\mathbf{I}+1)\mathbf{\Lambda}^+ = (\mathbf{I}+1)\mathbf{\Lambda} \text{ and } (\mathbf{I}-1)\mathbf{\Lambda}^- = (\mathbf{I}-1)\mathbf{\Lambda},$$

respectively. These formulas can be used to define Clifford parallels. A non-simple bivector which can be written as $\Xi^+ = (\mathbf{I}+1)\mathbf{\Lambda}$ where $\mathbf{\Lambda}$ is a line has the property that $\mathbf{I}\Xi^+ = \Xi^+$. We will call it a positive Clifford bivector. It defines a set of positive Clifford parallels via $(\mathbf{I}+1)\mathbf{\Lambda}^+ = \Xi^+$, which includes $\mathbf{\Lambda}$ and $\mathbf{\Lambda I}$. WeI will refer to these positive Clifford parallels as the Clifford parallels of the positive Clifford bivector $\mathbf{I}\Xi^+$. If $\mathbf{\Lambda}^+$ is a Clifford parallel of $\mathbf{I}\Xi^+$, then $\mathbf{\Lambda}^+ \mathbf{I}$ is also a Clifford parallel of Ξ^+. Similarly, a non-simple bivector which can be written as $\mathbf{I}\Xi^+ = (\mathbf{I} - 1)\ \mathbf{\Lambda}$ where Λ is a line has the property that $\mathbf{I}\Xi^- = \mathbf{I}\Xi^-$ and will be called a negative Clifford bivector. It defines a set of negative Clifford parallels via $(\mathbf{I}-1)\mathbf{\Lambda} - = \Xi^-$, which includes $\mathbf{\Lambda}$ and $-\mathbf{\Lambda I}$. We will refer to these negative Clifford parallels as the Clifford parallels of the negative Clifford bivector Ξ^-. If $\mathbf{\Lambda}^-$ is a Clifford parallel of Ξ^-, then $-\mathbf{\Lambda}^- \mathbf{I}$ is also a Clifford parallel of Ξ^-. The Clifford parallel $\mathbf{\Lambda}$ of a Clifford bivector Ξ which passes through a point $\mathbf{P}$ is given by,

$$\mathbf{\Lambda} = (\Xi \vee \mathbf{P})\mathbf{P}^{-1}.$$

It is either positive or negative depending on whether the Clifford bivector Ξ is positive or negative.

Most bivectors in E_3 are non-simple, but any non-simple bivector can be decomposed into a sum of two complementary lines called the axes of the non-simple bivector. Two lines Λ_1 and Λ_2 are called complementary if $\Lambda_1 \cdot \Lambda_2 = 0$ and $\Lambda_1 \times \Lambda_2 = 0$. Complementary lines are not only perpendicular but they also commute. For instance, Θ and $\Theta\mathbf{I}$, where Θ is a line, are complementary. For a non-simple Λ, assume $\Lambda = \Lambda_1 + \Lambda_2$, where Λ_1 and Λ_2 are complementary. Then,

$$\Lambda_1\Lambda_2 = \frac{1}{2}\Lambda \wedge \Lambda,$$
$$\Lambda_1^2 + \Lambda_2^2 = \Lambda \cdot \Lambda.$$

Squaring the first equation and using the second yields a quadratic equation for Λ_1^2 and Λ_2^2,

$$\Lambda_{1,2}^4 - (\Lambda \cdot \Lambda)_{1,2}^2 + \frac{1}{4}(\Lambda \vee \Lambda)^2 = 0.$$

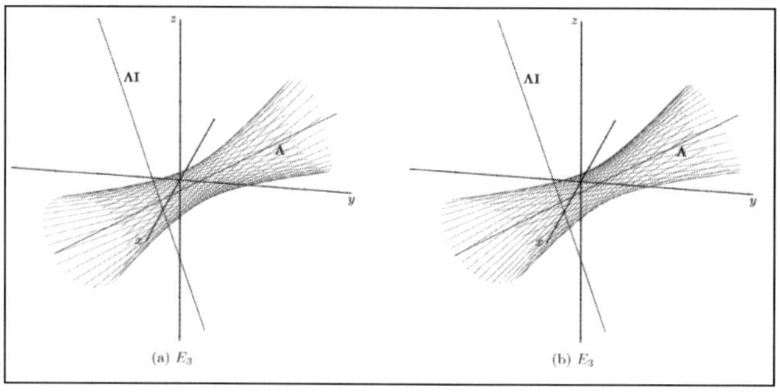

Clifford parallels of Λ.

If $(\Lambda \cdot \Lambda)^2 > (\Lambda \vee \Lambda)^2$, there are two distinct solutions given by,

$$\Lambda_{1,2}^4 = \frac{1}{2}\left(\Lambda \cdot \Lambda \pm \sqrt{(\Lambda \cdot \Lambda)^2 - (\Lambda \vee \Lambda)^2}\right).$$

For definiteness, we will assume Λ_1 uses the minus sign in $\Lambda_{1,2}^4 = \frac{1}{2}\left(\Lambda \cdot \Lambda \pm \sqrt{(\Lambda \cdot \Lambda)^2 - (\Lambda \vee \Lambda)^2}\right)$. and Λ_2 uses the plus sign, which implies $\|\Lambda_1\| > \|\Lambda_2\|$ since $\Lambda \cdot \Lambda$ is a negative scalar. We will call Λ_1 the larger axis and Λ_2 the smaller axis. If $(\Lambda \cdot \Lambda)^2 = (\Lambda \vee \Lambda)^2$ or equivalently $\Lambda \cdot \Lambda = \pm \Lambda \vee \Lambda$, then

$$\Lambda_1^2 = \Lambda_2^2 = \frac{1}{2}(\Lambda \cdot \Lambda).$$

These two cases exhaust all possibilities, since $(\Lambda \cdot \Lambda)^2 \geq (\Lambda \vee \Lambda)^2$ for any bivector Λ in E_3. Hence, if Λ_1^2 and Λ_2^2 are distinct, the axes of Λ are given by,

$$\Lambda_1 = \Lambda / \left(1 + \frac{1}{2}(\Lambda \wedge \Lambda)/\Lambda_1^2\right) \text{and } \Lambda_2 = \Lambda / \left(1 + \frac{1}{2}(\Lambda \wedge \Lambda)/\Lambda_2^2\right)$$

Note that $\Lambda_2 / \|\Lambda_2\| = \Lambda_1\mathbf{I} / \|\Lambda_1\|$ if $\Lambda \vee \Lambda < 0$ and $\Lambda_2 / \|\Lambda_2\| = \Lambda_1\mathbf{I} / \|\Lambda_1\|$ if $\Lambda \vee \Lambda > 0$, so the axes are at the distance $\frac{\pi}{2}$ from each other.

If $\Lambda_1^2 = \Lambda_2^2$, we get $\Lambda_{1,2}(1+\Lambda\wedge\Lambda/\Lambda\cdot\Lambda)=\Lambda$ but the multivector $1+\Lambda\wedge\Lambda/\Lambda\cdot\Lambda=1\pm\mathbf{I}$ is not invertible. In this case, the decomposition of the non-simple Λ into complementary lines is not unique. If $\Lambda \cdot \Lambda = \Lambda \vee \Lambda$, then

$$\Lambda = \lambda_1\mathbf{e}_{10} + \lambda_2\mathbf{e}_{20} + \lambda_3\mathbf{e}_{30} + \lambda_1\mathbf{e}_{23} + \lambda_2\mathbf{e}_{31} + \lambda_3\mathbf{e}_{12}$$

for some $\lambda_1, \lambda_2, \lambda_3 \in \mathbb{R}$ and Λ is a positive Clifford bivector. One possible decomposition is given by $\Lambda_1 = \lambda_1 e_{23} + \lambda_2 e_{31} + \lambda_3 e_{12}$ and $\Lambda_2 = \Lambda_1\mathbf{I} = \lambda_1 e_{10} + \lambda_2 e_{20} + \lambda e_{30}$. The bivector Λ can also be decomposed as $\Lambda = \Lambda_1^+ + \Lambda_1^+\mathbf{I}$, where Λ_1^+ is any positive Clifford parallel of Λ_1 (one needs to ensure that the weight of Λ_1^+ matches that of Λ_1). On the other hand, if $\Lambda \cdot \Lambda = -\Lambda \vee \Lambda$, then,

$$\Lambda = -\lambda_1 e_{10} - \lambda_2 e_{20} - \lambda_3 e_{30} + \lambda_1 e_{23} + \lambda_2 e_{31} + \lambda_3 e_{12}$$

for some $\lambda_1, \lambda_2, \lambda_3 \in \mathbb{R}$ and Λ is a negative Clifford bivector. A decomposition is given by $\Lambda_1 = \lambda_1\mathbf{e}_{23} + \lambda_2\mathbf{e}_{31} + \lambda_3\mathbf{e}_{12}$ and $\Lambda_2 = \Lambda_1\mathbf{I} = \lambda_1\mathbf{e}_{10} - \lambda_2\mathbf{e}_{20} - \lambda_3\mathbf{e}_{30}$. Other decompositions are given by $\Lambda = \Lambda_1^- - \Lambda_1^-\mathbf{I}$, where Λ_1^- is any negative Clifford parallel of Λ_1.

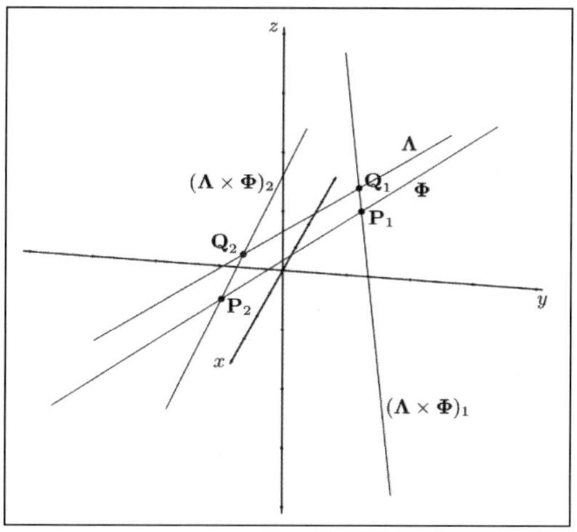

Commutator in E_3.

The commutator $\Lambda\times\Phi$ of two intersecting lines Λ and Φ is simple. It is a line which is perpendicular to both Λ and Φ and passes through Λ and Φ at the point where they intersect. On the other hand, if Λ and Φ are Clifford-parallel, then $\Lambda \times \Phi$ is generally not simple but the decomposition into axes is not unique. In fact, for any point on Λ (or on Φ), it is possible to find an axis which passes through the point. The separation r between Λ and Φ measured along the axes, i.e. the distance between the Clifford-parallel lines Λ and Φ, is constant and satisfies $\sin^2 r = |\Lambda \vee \Phi|$ and $\cos^2 r = |\Lambda \cdot \Phi|$ where Λ and Φ are assumed to be normalised.

If Λ and Φ do not intersect and are not Clifford-parallel, then $\Lambda \times \Phi$ is non-simple and can be decomposed uniquely into two axes, $(\Lambda \times \Phi)_1$ and $(\Lambda \times \Phi)_2$. The axes are perpendicular to both Λ and Φ and pass through both lines. Since the axes are perpendicular to both Λ and Φ, the separation between Λ and Φ measured along each axis is at a local minimum. If r1 and r2 denote these local minimal separations, then $r_1 < r_2$ and the distance r between the lines Λ and Φ is given by $r = r_1$, where r_1 corresponds to the separation along the larger axis $(\Lambda \times \Phi)_1$ and r_2 corresponds to the separation along the smaller axis $(\Lambda \times \Phi)_1$.

An example is shown in figure, where $\mathbf{\Lambda} = -\frac{3}{4}\mathbf{e}_{10} + \mathbf{e}_{20} - \frac{1}{2}\mathbf{e}_{30} - \mathbf{e}_{23} - \frac{5}{2}\mathbf{e}_{31} - 2\mathbf{e}_{12}$ and $\mathbf{\Phi} = \mathbf{e}_{10} + \frac{5}{3}\mathbf{e}_{20} - 2\mathbf{e}_{30} - \mathbf{e}_{23} + 3\mathbf{e}_{31} + 2\mathbf{e}_{12}$. We get $\sin r = \sin r_1 = \|\mathbf{P}_1 \vee \mathbf{Q}_1\| = \sqrt{\left(22 - 5\sqrt{17}\right)/59}$ for the normalised points $\mathbf{P}_1$ and $\mathbf{Q}_1$, where the axis $(\mathbf{\Lambda} \times \mathbf{\Phi})_1$ intersects the lines $\mathbf{\Lambda}$ and $\mathbf{\Phi}$ and $\sin r_2 = \|\mathbf{P}_2 \vee \mathbf{Q}_2\| = \sqrt{\left(22 + 5\sqrt{17}\right)59}$ for the normalised points $\mathbf{P}_2$ and $\mathbf{Q}_2$, where $(\mathbf{\Lambda} \times \mathbf{\Phi})_2$ intersects $\mathbf{\Lambda}$ and $\mathbf{\Phi}$.

The local minimal separations r_1 and r_2 satisfy $\cos r_1 \cos r_2 = |\mathbf{\Lambda} \cdot \mathbf{\Phi}|$ and $\sin r_1 \sin r_2 = |\mathbf{\Lambda} \vee \mathbf{\Phi}|$, provided that the lines are normalised. The rotation around the axis $(\mathbf{\Lambda} \times \mathbf{\Phi})_2$ causes the point $\mathbf{Q}_1$ to moves along $(\mathbf{\Lambda} \times \mathbf{\Phi})_1$. The distance $r = r_1$ can be thought of as the angle that the point $\mathbf{Q}_1$ needs to be rotated by in order to bring it to the point $\mathbf{P}_1$ (note that $\mathbf{Q}_1$ must be rotated in the correct direction to get r; if $\mathbf{Q}_1$ is rotated in the opposite direction one gets $\pi - r$ instead). Similarly, r_2 can be thought of as the angle required to bring $\mathbf{Q}_2$ to b_2 when the point $\mathbf{Q}_2$ is rotated around the axis $(\mathbf{\Lambda} \times \mathbf{\Phi})_1$. It is consistent with the angle between two planes, both of which pass through the axis $(\mathbf{\Lambda} \times \mathbf{\Phi})_1$ and one of which contains $\mathbf{\Lambda}$ while the other contains $\mathbf{\Phi}$.

By analogy with the Euclidean case, one can refer to r_2 as the angle between the lines $\mathbf{\Lambda}$ and $\mathbf{\Phi}$. Indeed, in the Euclidean case, the distance between two lines can be obtained by the rotation around the axis at infinity, i.e. translation, while the angle can be obtained by the rotation around the finite axis. In the elliptic case, both axes are finite and it is natural to associate the smaller separation with the distance and the larger separation with the angle between the lines. Since $r_2 \in \left[0, \frac{\pi}{2}\right]$, it does not take the orientation of the lines into account. It is more convenient to define the angle $\alpha \in [0, \pi]$ between the normalised lines $\mathbf{\Lambda}$ and $\mathbf{\Phi}$ by $\mathbf{\Lambda} \cdot \mathbf{\Phi} = -\cos r \cos \alpha$, where r is the distance between the lines, in order to take the orientation of the lines into account (r_2 may be called the unoriented angle and $r_2 = \alpha$ if $\alpha \le \frac{\pi}{2}$ or $r_2 = \pi - \alpha$ if $\alpha > \frac{\pi}{2}$). Hence, in general the geometric product of two normalised lines $\mathbf{\Lambda}$ and $\mathbf{\Phi}$ is given by,

$$\mathbf{\Lambda\Phi} = -\cos r \cos \alpha + \mathbf{\Lambda} \times \mathbf{\Phi} \pm \mathbf{I} \sin r \sin \alpha$$

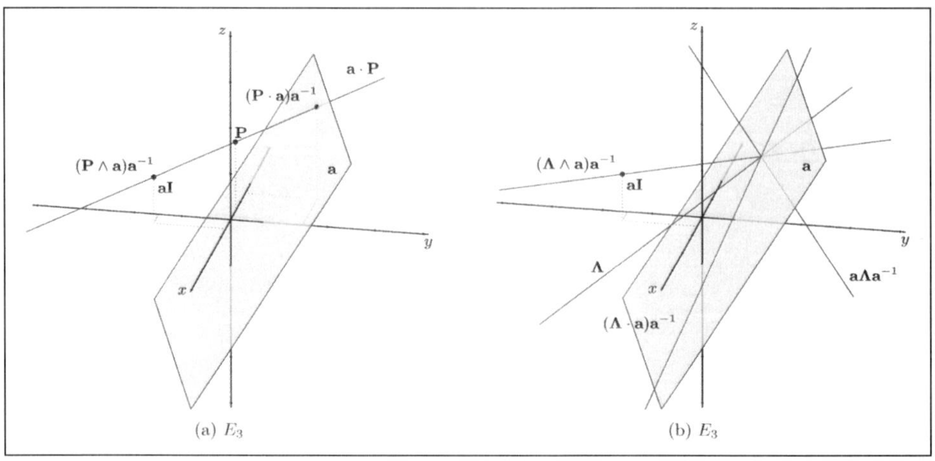

Projection, rejection and reflection in E_3.

where the sign in front of $\mathbf{I}$ depends on the relative orientation of the lines. The distance $r \in \left[0, \frac{\pi}{2}\right]$ between normalised lines $\mathbf{\Lambda}$ and $\mathbf{\Phi}$ can thus be determined from,

$$2\sin^2 r = 1 + v^2 - u^2 - \sqrt{\left(1 + v^2 - u^2\right)^2 - 4v^2},$$

where u = $\mathbf{\Lambda} \cdot \mathbf{\Phi}$ and $v = \mathbf{\Lambda} \vee \mathbf{\Phi}$. Once the distance is found, the angle α ∈ [0, π] is determined from cos α = −u/cos r. Since both distance and angular measures in E_3 are elliptic, there is not much difference between angles and distances as evidenced.

The usual formulas for projection and rejection apply in E_3. The projection of a geometric object dually represented by a blade B_l on a plane a is defined by $(B_l \cdot \mathbf{a})\,\mathbf{a}^{-1}$ and the rejection is defined by $(B_l \wedge \mathbf{a})\,\mathbf{a}^{-1}$. For instance the rejection of a line $\mathbf{\Lambda}$ by a is given by $(\mathbf{\Lambda} \wedge \mathbf{a})\,\mathbf{a}^{-1}$. The rejections pass through the polar point $\mathbf{aI}$ of the plane a, and the projections lie on a. The rejection by **a** is zero if Bl lies on the plane a, and the projection on a is zero if B_l passes through the polar point $\mathbf{aI}$ of a, in which case B_l is perpendicular to **a**. An example of the projection on and rejection by the plane $\mathbf{a} = \mathbf{e}_0 + \frac{1}{2}\mathbf{e}_1 - \frac{3}{2}\mathbf{e}_2 + \mathbf{e}_3$ is shown in figure for the point $\boldsymbol{P} = \mathbf{e}_{123} - \frac{1}{2}\mathbf{e}_{320} + \frac{3}{2}\mathbf{e}_{210}$ and the line $\mathbf{\Lambda} = \mathbf{e}_{10} + \mathbf{e}_{20} - 3\mathbf{e}_{23} + 3\mathbf{e}_{31} + 2\mathbf{e}_{12}$.

The projection of B_l on a point P is defined by $(B_l \cdot \mathbf{P})\,\mathbf{P}^{-1}$ and the rejection is defined by $(B_l \wedge \mathbf{P})\,\mathbf{P}^{-1}$ for planes (l = 1) and by $(B_l \times \mathbf{P})\,\mathbf{P}^{-1}$ for lines and points (l = 2, 3), in accord with the decomposition of the relevant geometric products. For instance, the rejection of a line $\mathbf{\Lambda}$ by **P** is given by $(\mathbf{\Lambda} \times \mathbf{P})\,\mathbf{P}^{-1}$. The rejections lie on the plane $\mathbf{PI}$, whose polar point is **P**, and the projections pass through the point **P**. The projection of B_l on **P** is zero if B_l lies on **PI**, and the rejection of B_l by **P** is zero if B_l passes through **P**. The configurations are in some sense dual to those arising in projection on and rejection by a plane.

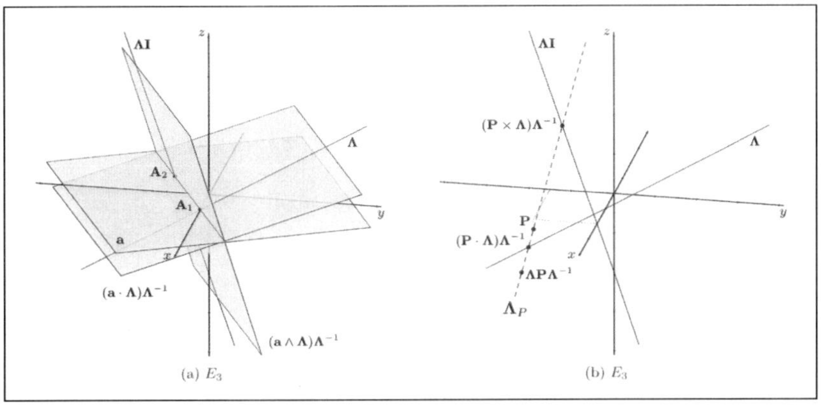

Projection, rejection and reflection in E_3.

The projection of B_l on a line $\mathbf{\Lambda}$ is defined by $(B_l \cdot \mathbf{\Lambda})\,\mathbf{\Lambda}^{-1}$ for points and planes (l = 1, 3). The rejection of a plane a by $\mathbf{\Lambda}$ is defined by $(\mathbf{a} \wedge \mathbf{\Lambda})\,\mathbf{\Lambda}^{-1}$, and the rejection of a point **P** by $\mathbf{\Lambda}$ is defined by $(\mathbf{P} \times \mathbf{\Lambda})\,\mathbf{\Lambda}^{-1}$. An example of the projection on and rejection by the line $\mathbf{\Lambda} = -\frac{1}{3}\mathbf{e}_{20} + \mathbf{e}_{30} + \mathbf{e}_{23} - \mathbf{e}_{31} - \frac{1}{3}\mathbf{e}_{12}$ is shown in figure for the plane $\mathbf{a} = \mathbf{e}_3$ and the point $\mathbf{P} = \boldsymbol{e}_{123} + 2\boldsymbol{e}_{320} - \frac{3}{2}\boldsymbol{e}_{130} - \frac{1}{4}\boldsymbol{e}_{210}$. In figure, the points Λ_1 and Λ_2 are located at the intersect of the plane **a** with the lines Λ and ΛI, respectively. The projection $(\mathbf{a} \cdot \Lambda)\Lambda^{-1}$ contains Λ and the rejection $(\mathbf{a} \wedge \Lambda)\Lambda^{-1}$ contains ΛI. The projection $(\mathbf{P} \cdot \Lambda)\Lambda^{-1}$ and rejection $(\mathbf{P} \times \Lambda)\Lambda^{-1}$ of the point **P** by the line Λ both lie on the line $\Lambda_p = (\Lambda \cdot \mathbf{P}) \wedge (\Lambda \vee \mathbf{P})$, which passes through **P** and is perpendicular to Λ; $\Lambda \mathbf{P}$ is also perpendicular to ΛI. Note that Λ_p can also be written as $\Lambda_p = \frac{1}{2}\mathbf{P} \vee (\Lambda \mathbf{P} \Lambda^{-1})$ if Λ is normalised.

The geometric product $\Phi\,\Lambda$ of two lines consists of three terms which can be split into two components for the projection and rejection in two different ways. In general, the commutator $\Phi \times \Lambda$ is not simple, so it needs to be decomposed into two axes. Then the larger axis $(\Phi \times \Lambda)_1$ can be combined with $\Phi \cdot \Lambda$ and the smaller axis $(\Phi \times \Lambda)_2$ with $\Phi \wedge \Lambda$, which gives the first kind of projection and rejection:

$$\mathrm{proj}_1(\Phi;\ \Lambda) = \left(\Phi \cdot \Lambda + (\Phi \times \Lambda)_1\right)\Lambda^{-1}$$
$$\mathrm{rej}_1(\Phi;\ \Lambda) = \left((\Phi \times \Lambda)_2 + \Phi \wedge \Lambda\right)\Lambda^{-1}$$

If $(\Phi \times \Lambda)_1$ is combined with $\Phi \wedge \Lambda$ and $(\Phi \times \Lambda)_2$ with $\Phi \cdot \Lambda$, one gets the second kind of projection and rejection:

$$\mathrm{proj}_2(\Phi;\ \Lambda) = \left(\Phi \cdot \Lambda + (\Phi \times \Lambda)_2\right)\Lambda^{-1},$$
$$\mathrm{rej}_2(\Phi;\ \Lambda) = \left((\Phi \times \Lambda)_1 + \Phi \wedge \Lambda\right)\Lambda^{-1}.$$

These projections and rejections share some of their basic properties with those in Euclidean space. For instance, $\mathrm{proj}_1(\Phi;\Lambda)$ is a line which passes through Λ and is at the same angle to Λ as Φ, and $\mathrm{rej}_2(\Phi;\Lambda)$ is a line which is at the same distance from Λ as Φ and is perpendicular to Λ.

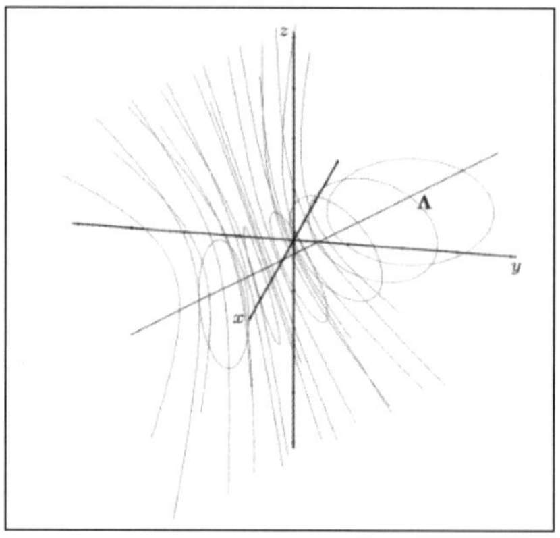

Rotation in E_3.

The top-down reflection of B_l in A_k is given by $(-1)^{kl} A_k B_l A_k^{-1}$ and the bottom-up reflection is given by $(-1)^{k(l-1)} A_k B_l A_k^{-1}$. The top-down and bottom-up reflections in lines are identical since k = 2 for lines. For instance, the reflection of a point **P** in a line Λ is given by $\Lambda \mathbf{P} \Lambda^{-1}$, and the top-down reflection of Λ in a plane a is given by $\mathbf{a} \Lambda \mathbf{a}^{-1}$. Note that the points **P** and $\Lambda \mathbf{P} \Lambda^{-1}$ are at the same distance from Λ, and the line $\mathbf{a} \Lambda \mathbf{a}^{-1}$ is at the same angle to the plane a as Λ.

Spinors in E_3 are defined in the usual way as multivectors that can be written as the product of an even number of planes which square to unity. They are even and each spinor satisfies $S\tilde{S}$ = 1. Any spinor in E_3 can also be written as $S = e^A$ for some bivector A. Spinors form a Lie group whose Lie algebra consists of bivectors with the commutator used as the product in the algebra.

Any proper motion in E_3 can be obtained as an action of a spinor. One encounters the following proper motions. The action SA_kS^{-1} of a spinor,

$$S = e^{-\frac{1}{2}(\alpha+\beta\mathbf{I})\Lambda},$$

Where $\alpha,\ \beta \in \mathbb{R}, \alpha \neq \beta$, and $\mathbf{\Lambda}$ is a normalised line, on the geometric object dually represented by a blade Ak yields a double rotation of A_k around $\mathbf{\Lambda}$ by the angle α and around $\mathbf{\Lambda I}$ by β. Under the double rotation, the axes $\mathbf{\Lambda}$ and $\mathbf{\Lambda I}$ of the double rotation are invariant, which implies that any point on $\mathbf{\Lambda}$ or $\mathbf{\Lambda I}$ stays on the line. Double rotation in E_3 corresponds to a proper motion in E_3 which consists of a simple rotation around an axis and a translation along the same axis. If β = 0, a simple rotation around $\mathbf{\Lambda}$ is obtained. Some trajectories generated by the rotation around the line $\mathbf{\Lambda}$ are shown in figure.

If one substitutes α = β in the above, the spinor can be written as,

$$S = e^{-\frac{1}{2}\beta\Xi^+}$$

where $\Xi^+ = (\mathbf{I}+1)\mathbf{\Lambda}$ is a positive Clifford bivector, and the resulting proper motion occurs along the Clifford parallels associated with $\Xi^+ = (\mathbf{I}+1)\mathbf{\Lambda}$. If, on the other hand, α = −β, then:

$$S = e^{-\frac{1}{2}\beta\Xi^+}$$

where $\Xi^- = (\mathbf{I}-1)\mathbf{\Lambda}$ is a negative Clifford bivector, and the proper motion is along the Clifford parallels of Ξ^-. The proper motions which result from the action SA_kS^{-1} of S defined by $S = e^{-\frac{1}{2}\beta\Xi^+}$ and $S = e^{-\frac{1}{2}\beta\Xi^+}$ are called Clifford translations. The Clifford translation of a point **P** generated by a Clifford bivector Ξ can also be written as:

$$\mathbf{P} \to \mathbf{P}\cos\beta + \mathbf{P}\times\Xi\ \sin\beta$$

by expressing Ξ in terms of the Clifford parallel that passes through **P** and simplifying the action SPS^{-1} of the spinor $S - e^{-\frac{1}{2}\beta\Xi}$. Clifford translations can also be represented by quaternion multiplication. If a point $\mathbf{P} = w\mathbf{e}_{123} + x\mathbf{e}_{320} + y\mathbf{e}_{130} + z\mathbf{e}_{210}$ is normalised and $\mathbf{\Lambda} = \lambda_1\mathbf{e}_{23} + \lambda_2\mathbf{e}_{31} + \lambda_3\mathbf{e}_{12}$, then the Clifford translation of **P** by β generated by the positive Clifford bivector $\Xi = \Xi^+ = (\mathbf{I}+1)\mathbf{\Lambda}$ is also given by $p \mapsto pq$, where $p = w + x\boldsymbol{i} + y\boldsymbol{j} + z\boldsymbol{k}$ and $q = \cos\beta - (\lambda_1 i + \lambda_2 j + \lambda_3 k)\sin\beta$ are the relevant quaternions. So, a Clifford translation generated by Ξ^+ can also be referred to as a right Clifford translation. For the negative Clifford bivector $\Xi = \Xi^- = (\mathbf{I} - 1)\ \mathbf{\Lambda}$, $p \mapsto qp$ yields the same Clifford translation as that generated by Ξ^-. It can be called a left Clifford translation.

Quasi-elliptic Space

A projective n-space whose projective metric is defined by an absolute consisting of an imaginary cone (the absolute cone Q_0) with an $(n-m-1)$-vertex (the absolute plane T_0) together with an imaginary $(n-m-2)$-quadric Q_1 on this $(n-m-1)$-plane (the absolute quadric Q_1); it is denoted by the symbol S_n^m, $m<n$. A quasi-elliptic space is of more general projective type in comparison with a Euclidean space and a co-Euclidean space; the metrics of the latter are obtained from those of the former. A quasi-elliptic space is a particular case of a semi-elliptic space. For $m=0$, the absolute cone is a pair of coincident $(n-1)$-planes coinciding with the $(n-1)$-absolute plane T_0, while the absolute coincides with the absolute of Euclidean n-space. For $m=1$, the cone Q_0 is a cone with a point vertex and the absolute in this case is the same as that of the co-Euclidean n-space. When $m=1$, the cone Q_0 is a pair of imaginary $(n-1)$-planes. In particular, the cone Q_0 of the quasi-elliptic three-space S_3^1 is a pair of imaginary two-planes, the line (the 1-plane) T_0 is the real line of their intersection, while the quadric Q_1 is a pair of imaginary points on T_0.

The distance δ between two points X and Y is defined in case the line XY does not intersect the $(n-m-1)$-plane T_0 by the relation,

$$\cos^2\frac{\delta}{\rho}=\frac{\left(x^0E_0y^0\right)^2}{\left(x^0E_0x^0\right)\left(y^0E_0y^0\right)},$$

$$x^0=\left(x^a,a\le m\right),\quad y^0=\left(y^b,b\le m\right)$$

are the vectors of the points X and Y, E_0 is the linear operator defining the scalar product in the space of these vectors and ρ is a real number; in case XY intersects T_0, the distance d between these points is defined by means of the distance between the vectors of the points X and Y:

$$x=y^1-x^1,$$

$$x^1=\left(x^a,a>m\right),\quad y^1=\left(y^b,b>m\right),$$

$$d^2=aE_1a,$$

where E_1 is the linear operator defining the scalar product in the space of these vectors.

The angle between two planes whose $(n-2)$-plane of intersection does not intersect the $(n-m-1)$-plane T_0 is defined as the (normalized) distance between the corresponding points in the dual quasi-elliptic space S_n^{n-m-1}, in which the coordinates are numerically equal or proportional to the projective coordinates of the planes in S_n^m. If the $(n-2)$-plane of intersection of two given planes intersects the $(n-m-1)$-plane T_0, then the angle between the planes is in this case again defined by the numerical distance. When $n=2$ the angles between the planes are the angles between the lines.

The motions of the quasi-elliptic space S_n^m are the collineations of this space that take the cone Q_0 into the plane T_0 and the quadric Q_1 into itself. The group of motions is a Lie group and the motions are described by orthogonal operators. In the quasi-elliptic space S_{2m+1}^m, which is self-dual, co-motions are defined as the correlations that take each pair of points into two 2m-planes the angle between which is proportional to the distance between the points, and each pair of 2m-planes into two points the distance between which is proportional to the angle between the planes. The motions and co-motions of S^m_{2m+1} form a group, which is a Lie group. The geometry of the 2-plane S^0_2 is Euclidean geometry, while the geometry of the 2 -plane S^1_2 is the same as that of the co-Euclidean plane.

The geometry of the 3-space S^1_3 is defined by an elliptic projective metric on lines that is co-Euclidean on planes and Euclidean in bundles of planes. The geometry of the 3-space S^0_3 is Euclidean, while the geometry of the 3-space S^2_3 is the same as that of the co-Euclidean 3-space. The space S^1_3 with radius of curvature 1/2 is isometric to the connected group of motions of the Euclidean 2-space with a specially introduced metric. The connected group of motions of the quasi-Euclidean space S^1_3 is isomorphic to the direct product of two connected groups of motions of the Euclidean 2-plane.

Spherical Geometry

Spherical geometry is the study of geometric objects located on the surface of a sphere. Spherical geometry works similarly to Euclidean geometry in that there still exist points, lines, and angles. For instance, a "line" between two points on a sphere is actually a great circle of the sphere, which is also the projection of a line in three-dimensional space onto the sphere.

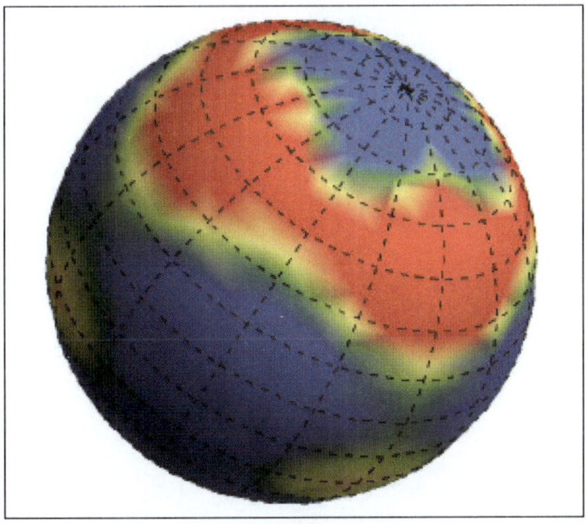

However, it differs from typical Euclidean geometry in several substantial ways.

Properties of Spherical Geometry

In spherical geometry, there are no parallel lines. In fact, there are no straight lines. Therefore, in spherical geometry, a great circle is comparable to a line. There are no straight lines in spherical geometry. Instead, the shortest distance from one point to the next lying on a sphere is along the arc of a great circle. On a sphere, the angle between two curved arcs is measured by the angle formed from the intersection of the lines lying tangent to the two arcs.

When three curved arcs intersect one another, a spherical triangle is formed. A spherical triangle is any three-sided region enclosed by sides that are the arcs of great circles. Spherical triangles can have angle sums that range between one hundred eighty degrees and five hundred forty degrees. Spherical triangles can have angle sums larger than the usual one hundred eighty degrees found in a triangle because lines connecting points have a slight curve to them. Even so, spherical triangles can have ninety-degree angles just like triangles in the plane.

Differing from Euclidean geometry, two spherical triangles are not only similar, but congruent if they share the same angles. If one of the angles of a spherical triangle is a right angle, the triangle is known as a spherical right triangle, and a Spherical Pythagorean Theorem exists. In a spherical right triangle, let C denote the length of the side opposite the right angle. Let A and B denote the lengths of the other two sides. Let R denote the radius of the sphere. Then, we can use the following formula: $\cos(C/R) = \cos(A/R) * \cos(B/R)$.

To calculate the area of a spherical triangle on a unit sphere, one must sum the angles of the spherical triangle (in radians) and subtract pi. For example, say a spherical triangle had two right angles and one forty-five degree angle. To find the area of the spherical triangle, restate the angles given in degrees to angles in radians. Thus, we are working with a spherical triangle with two pi/2 angles and one pi/4 angle. Add the three angles together $(pi/2 + pi/2 + pi/4)$. Thus, the angles total 5pi/4. Subtract pi from 5pi/4 to find the area of the spherical triangle. Therefore, we have a total area of pi/4 units squared.

It has been established that a great circle is formed from a plane that intersects a sphere through its center. In analyzing two great circles that lie in a sphere, the two planes that form the great circles must intersect in a line, which in turn intersects the sphere at two distinct points. Therefore, two great circles intersect each other at two antipodal points. The planes also divide the sphere into four parts. These regions are known as lunes. A lune is a two-sided polygon lying on a sphere.

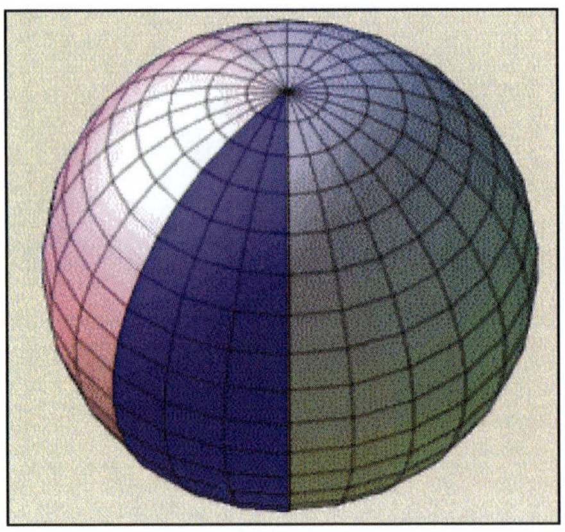

To find the area of a lune, one must know the lunar angle and the formula $2R^2$ (lunar angle), where R represents the radius of the sphere. If the lunar angle is not given, find p(pi)/q, where p represents the union of the lunes and q is the number of equal lunes drawn in a hemisphere. Knowledge about a lune's area can be helpful in solving the area of the sphere. For example, say the area of a lune is thirty-two centimeters squared, and we want to find the area of the sphere where the lune is present. The area of a sphere is $4piR^2$, where R represents the radius. Since we know the formula for the area of a lune, we can solve for the radius, the unknown variable. Then we can use the radius in the formula for the area of the sphere to solve for the area of the sphere. After a few calculations, we can see that the radius is four centimeters. We then can square this and multiply the new result by 4pi. Thus, we have the sphere's area equal to 64pi centimeters squared.

The shortest distance between two points on a sphere is along the arc of a great circle. In spherical geometry, there exists a formula to find this distance. Given two points A and B, first locate the angle (measured in radians) that is created between the two points and the center of the sphere. Then, multiply by the radius (R) of the sphere, giving the formula d(A, B)= Ra, where a represents the angle formed between points A and B. For example, to find the distance between two points A and B with an angle of pi/4 between them on a sphere having a radius of 4, we multiply pi/4 by the radius length of 4 units to get a spherical distance of pi.

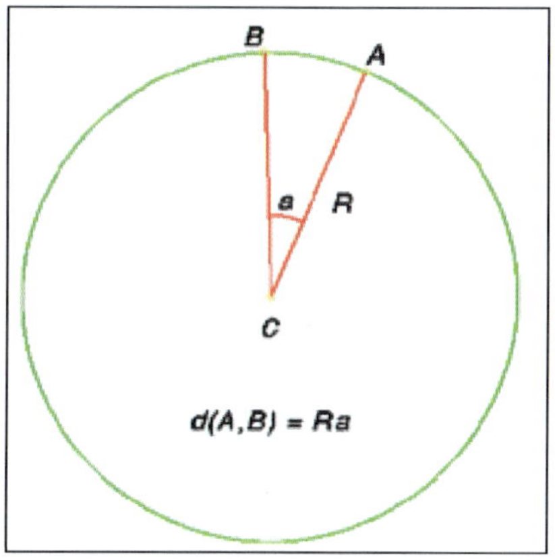

Relation to Euclid's Postulates

Spherical geometry obeys two of Euclid's postulates: the second postulate ("to produce [extend] a finite straight line continuously in a straight line") and the fourth postulate ("that all right angles are equal to one another"). However, it violates the other three: contrary to the first postulate, there is not a unique shortest route between any two points (antipodal points such as the north and south poles on a spherical globe are counterexamples); contrary to the third postulate, a sphere does not contain circles of arbitrarily great radius; and contrary to the fifth (parallel) postulate, there is no point through which a line can be drawn that never intersects a given line.

A statement that is equivalent to the parallel postulate is that there exists a triangle whose angles add up to 180°. Since spherical geometry violates the parallel postulate, there exists no such triangle on the surface of a sphere. The sum of the angles of a triangle on a sphere is $180°(1 + 4f)$, where f is the fraction of the sphere's surface that is enclosed by the triangle. For any positive value of f, this exceeds 180°.

Hyperbolic Geometry

Hyperbolic geometry is a non-Euclidean geometry that rejects the validity of Euclid's fifth, the "parallel," postulate. Simply stated, this Euclidean postulate is: through a point not on a given line there is exactly one line parallel to the given line. In hyperbolic geometry, through a point not on a given line there are at least two lines parallel to the given line. The tenets of hyperbolic geometry, however, admit the other four Euclidean postulates.

Although many of the theorems of hyperbolic geometry are identical to those of Euclidean, others differ. In Euclidean geometry, for example, two parallel lines are taken to be everywhere equidistant. In hyperbolic geometry, two parallel lines are taken to converge in one direction and diverge in the other. In Euclidean, the sum of the angles in a triangle is equal to two right angles; in hyperbolic, the sum is less than two right angles. In Euclidean, polygons of differing areas can be similar; and in hyperbolic, similar polygons of differing areas do not exist.

Hypercycles and Horocycles

There is a curve peculiar to hyperbolic geometry, called the horocycle. Consider two limiting parallel lines, ℓ and m, with a common direction, say Ω. Let P be a point on one of these lines $p \in \ell$. If there exists a point $Q \in m$ such that the singly asymptotic triangle, $\Delta PQ\Omega$, has the property that:

$$\angle PQ\Omega \cong \angle QP\Omega$$

then we say that Q corresponds to P. If the singly asymptotic triangle $\Delta PQ\Omega$ has the above property we shall say that it is equiangular. Note that it is obvious from the definition that if Q corresponds to P, then P corresponds to Q. The points P and Q are called a pair of corresponding points.

Theorem: If points P and Q lie on two limiting parallel lines in the direction of the ideal point, Ω, they are corresponding points on these lines if and only if the perpendicular bisector of PQ is limiting parallel to the lines in the direction of Ω.

Theorem: Given any two limiting parallel lines, there exists a line each of whose points is equidistant from them. The line is limiting parallel to them in their common direction.

Proof: Let ℓ and m be limiting parallel lines with common direction Ω. Let $A \in \ell$ and $B \in m$. The bisector of $\angle BA\Omega$ in the singly asymptotic triangle $\Delta AB\Omega$ meets side BΩ in a point X and the bisector of $\angle AB\Omega$ meets side AX of the triangle ΔABX in a point C. Thus the bisectors of the angles of the singly asymptotic triangle ABΩ meet in a point C. Drop perpendiculars from C to each of ℓ and m, say P and Q, respectively. By Hypothesis-Angle $\Delta CAP \cong \Delta CAM$ (M is the midpoint of AB) and $\Delta CBQ \cong \Delta CBM$. Thus, $CP \cong CM \cong CQ$. Thus, by *SAS* for singly asymptotic triangles, we have that,

$$\Delta CP\Omega \cong \Delta CQ\Omega$$

and thus the angles at C are congruent. Now, consider the line $\overleftrightarrow{C\Omega}$ and let F be any point on it other than C. By *SAS* we have $\Delta CPF \cong \Delta CQF$. If S and T are the feet of F in ℓ and m, then we get that $\Delta PSF \cong \Delta QTF$ and $FS \cong FT$. Thus, every point on the line CΩ is equidistant from ℓ and m.

This line is called the equidistant line.

Theorem: Given any point on one of two limiting parallel lines, there is a unique point on the other which corresponds to it.

Theorem: If three points P, Q, and R lie on three parallels in the same direction so that P and Q are corresponding points on their parallels and Q and R are corresponding points on theirs, then P, Q, and R are noncollinear.

Theorem: If three points P, Q, and R lie on three parallels in the same direction so that P and Q are corresponding points on their parallels and Q and R are corresponding points on theirs, then P and R are corresponding points on their parallels.

Consider any line ℓ, any point $P \in \ell$, and an ideal point in one direction of ℓ, say Ω. On each line parallel to ℓ in the direction Ω there is a unique point Q that corresponds to P. The set consisting of P and all such points Q is called a horocycle, or, more precisely, the horocycle determined by ℓ, P, and Ω. The lines parallel to ℓ in the direction Ω, together with ℓ, are called the radii of the horocycle. Since ℓ may be denoted by PΩ, we may regard the horocycle as determined simply by P and Ω, and hence call it the horocycle through P with direction Ω, or in symbols, the horocycle (P, Ω).

All the points of this horocycle are mutually corresponding points by Theorem, so the horocycle is equally well determined by any one of them and Ω. In other words if Q is any point of horocycle (P, Ω) other than P, then horocycle (Q, Ω) is the same as horocycle (P, Ω). If, however, P' is any point of ℓ other than P, then horocycle (P', Ω) is different from horocycle (P, Ω), even though they have the same direction and the same radii. Such horocycles, having the same direction and the same radii, are called codirectional horocycles.

There are analogies between horocycles and circles.

Lemma: There is a unique horocycle with a given direction which passes through a given point. (There is a unique circle with a given center which passes through a given point.)

Lemma: Two codirectional horocycles have no common point. (Two concentric circles have no common point.)

Lemma: A unique radius is associated with each point of a horocycle. (A unique radius is associated with each point of a circle.)

A tangent to a horocycle at a point on the horocycle is defined to be the line through the point which is perpendicular to the radius associated with the point.

No line can meet a horocycle in more than two points. This is a consequence of the fact that no three points of a horocycle are collinear inasmuch as it is a set of mutually corresponding points.

Theorem: The tangent at any point A of a horocycle meets the horocycle only in A. Every other line through A except the radius meets the horocycle in one further point B. If α is the acute angle between this line and the radius, then d(A, B) is twice the segment which corresponds to α as angle of parallelism.

Proof: Let t be the tangent to the horocycle at A and let Ω be the direction of the horocycle. If t met the horocycle in another point B, we would have a singly asymptotic triangle with two right angles, since A and B are corresponding points. In fact the entire horocycle, except for A, lies on the same side of t, namely, the side containing the ray AΩ.

Let k be any line through A other than the tangent or radius. We need to show that k meets the horocycle in some other point. Let α be the acute angle between k and the ray AΩ. Let C be the point of k, on the side of t containing the horocycle, such that AC is a segment corresponding to α as angle of parallelism. The line perpendicular to k at C is then parallel to AΩ in the direction Ω. Let B be the point of k such that C is the midpoint of AB. The singly asymptotic triangles ΔACΩ and ΔBCΩ are congruent. Hence ∠CBΩ = α, B corresponds to A, and B ∈ (A, Ω).

A chord of a horocycle is a segment joining two points of the horocycle.

Theorem: The line which bisects a chord of a horocycle at right angles is a radius of the horocycle.

We can visualize a horocycle in the Poincare model as follows. Let ℓ be the diameter of the unit disk whose interior represents the hyperbolic plane, and let O be the origin. It is a fact that the hyperbolic circle with hyperbolic center P is represented by a Euclidean circle whose Euclidean center R lies between P and A.

As P recedes from A towards the ideal point Ω, R is pulled up to the Euclidean midpoint of ΩA, so that the horocycle (A, Ω) is a Euclidean circle tangent to the unit disk at Ω and tangent to ℓ at A. It can be shown that all horocycles are represented in the Poincare model by Euclidean circles inside the unit disk and tangent to boundary circle. For the Poincare upper half plane model, our horocycles will be circles that are tangent to the x-axis.

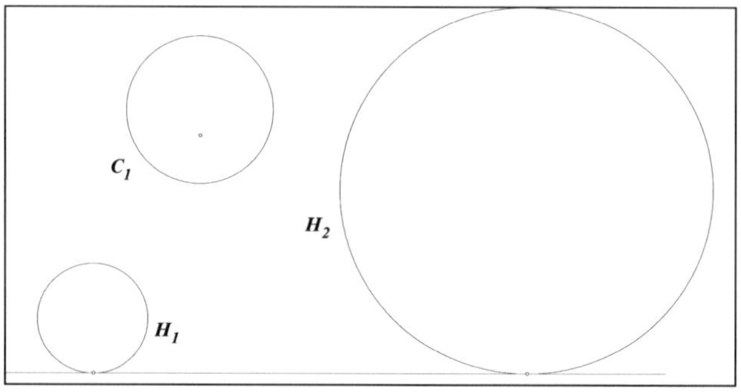

H_1 and H_2 are horocycles in the Poincare model, C_1 is a circle.

Another curve found specifically in the hyperbolic plane and nowhere else is the equidistant curve, or hypercycle. Given a line ℓ and a point P not on ℓ, consider the set of all points Q on one side of ℓ so that the perpendicular distance from Q to ℓ is the same as the perpendicular distance from P to ℓ.

The line ℓ is called the axis, or base line, and the common length of the perpendicular segments is called the distance. The perpendicular segments defining the hypercycle are called its radii. The following statements about hypercycles are analogous to statements about regular Euclidean circles:

1. Hypercycles with equal distances are congruent, those with unequal distances are not. (Circles with equal radii are congruent, those with unequal radii are not.)

2. A line cannot cut a hypercycle in more than two points.

3. If a line cuts a hypercycle in one point, it will cut it in a second unless it is tangent to the curve or parallel to it base line.

4. A tangent line to a hypercycle is defined to be the line perpendicular to the radius at that point. Since the tangent line and the base line have a common perpendicular, they must be hyperparallel. This perpendicular segment is the shortest distance between the two lines. Thus, each point on the tangent line must be at a greater perpendicular distance from the

base line than the corresponding point on the hypercycle. Thus, the hypercycle can intersect the hypercycle in only one point. A line perpendicular to a chord of a hypercycle at its midpoint is a radius and it bisects the arc subtended by the chord.

5. Two hypercycles intersect in at most two points.

6. No three points of a hypercycle are collinear.

In the Poincare disk model let Ω and Λ be the ideal end points of ℓ. It can be shown that the hypercycle to ℓ through Ω is represented by the arc of the Euclidean circle passing through A, B, and Ω. This curve is orthogonal to all Poincare lines perpendicular to the line *l*. In the Poincare upper half plane model, the hypercycle will be represented by an arc of a Euclidean circle passing through A, B, and Ω.

In the Poincare disk model, $\mathscr{D}$, a Euclidean circle represents:

1. A hyperbolic circle if it is entirely inside the unit disk;
2. A horocycle if it is inside the unit disk except for one point where it is tangent to the unit disk;
3. An equidistant curve if it cuts the unit disk non-orthogonally in two points;
4. A hyperbolic line if it cuts the unit disk orthogonally.

A similar situation is true for the Poincare upper half plane model, $\mathscr{H}$.

It follows that in the hyperbolic plane three non-collinear points lie either on a circle, a horocycle, or a hypercycle accordingly, as the perpendicular bisectors of the triangle are concurrent in an ordinary point, an ideal point, or an ultra-ideal point.

Congruency of Horocycles

Consider a first horocycle H_1 with point p on the circle S^1_∞ and a second horocycle H_2 with point q on the circle S^1_∞. We need to find a transformation in $\mathcal{H}$ mapping H_1 to H_2. The transformation $T_1(z) = e^{-\arg(p)}z$ maps H_1 to a horocycle with the same radius as H_1 (when considered a Euclidean circle) and point (1, 0) on S^1_∞. The transformation $T_2(z) = e^{-\arg(q)}z$ maps H_2 to a horocycle with the same radius as H_2 (when considered a Euclidean circle) and point (1, 0) on S^1_∞. Notice that T_1, $T_2 \in \mathcal{H}$.

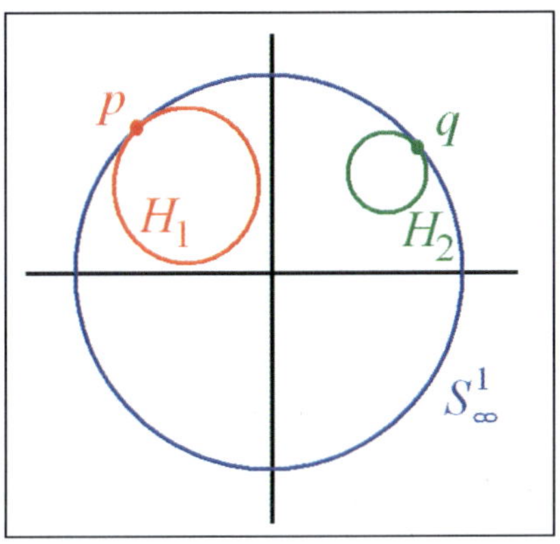

Let a be the leftmost point on $T_1(H_1)$ and b the leftmost point on $T_2(H_2)$. Then both a and b are real. Now we map a to b with a hyperbolic transformation of the form $T_3(z)=e^{i\theta}\frac{z-\alpha}{\bar{\alpha}z-1}$. We find that we can use $T_3(z)=\frac{(1-ab)z-(a-b)}{(1-ab)-(a-b)z}$ (based on the fact that a and b are real, we can deduce that α is real). Notice that $T_3 \in \mathcal{H}$ and $T_3(1)=1$.

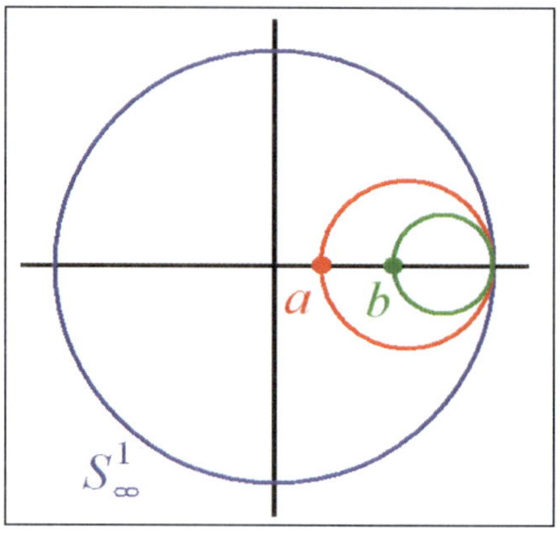

We know that a Möbius transformation is determined by it's behavior on three points. So if we can show that a third point of $T_1(H_1)$ is mapped to $T_2(H_2)$, then this would establish that T_3 maps $T_1(H_1)$ to $T_2(H_2)$. The uppermost point on $T_1(H_1)$ is $\frac{1+a}{2}+i\frac{1-a}{2}$. If we apply T_3 to this point we find that:

$$T_3\left(\frac{1+a}{2}+i\frac{1-a}{2}\right)=\frac{a^2b+2ab+1+b^2+3b}{a^2+2a+2+2b+b^2}+i\frac{-ab^2+a-b^2+1}{a^2+2a+2+2b+b^2}.$$

The (Euclidean) center of $T_2(H_2)$ is $\frac{1+b}{2}+i0$ and the (Euclidean) radius is —— and so we consider:

$$\left(\frac{a^2b+2ab+1+b^2+3b}{a^2+2a+2+2b+b^2}-\left(\frac{1+b}{2}\right)\right)^2+\left(\frac{-ab^2+a-b^2+1}{a^2+2a+2+2b+b^2}\right)^2$$

and find that this reduces to $\left(\frac{1-b}{2}\right)^2$. Therefore the uppermost point of $T_1(H_1)$ is mapped to $T_2(H_2)$ and we know that $T_3(T_1(H_1))=T_2(H_2)$. So we have $T_2^{-1}(T_3(T_1(H_1)))=H_2$.

Apeirogon

The apeirogon is an extension of the definition of regular polygon to a figure with an infinite number of sides. Its Schläfli symbol is $\{\infty\}$.

The apeirogon can produce a regular tiling on the hyperbolic plane. This is achieved by letting each edge of a regular polygon have length s, and each internal angle of the polygon be θ. Then construct triangle ABC, where C is the midpoint of an edge, B is an adjacent vertex, and A is the center of the polygon. This is a right triangle, with the right angle at C. But the length of side a is s/2, and the angle B is $\theta / 2$. The length of side c, the radius of the circumscribed circle, can the be determined using the standard formula for a right triangle on a surface that has a constant curvature of −1,

$$\begin{aligned}\tanh c &= \tanh a \sec B \\ &= \tanh\left(\frac{1}{2}s\right)\sec\left(\frac{1}{2}\theta\right).\end{aligned}$$

The value of $\sec(\theta / 2)$ is never less than one, while the value of $\tanh(s / 2)$ increases from zero to one as s increases. Only for small values of s is $\tanh c$ less than one, which it has to be for any real value of c. Thus, by making s large enough, the figure has an infinite number of sides and is an apeirogon. If $\tanh(s / 2)\sec(\theta / 2) = 1$, the figure is inscribed in a horocycle. If that value is greater than one, the figure is inscribed in a hypercycle or equidistant curve.

To tile the hyperbolic plane with apeirogons, select a Schläfli symbol $\{\infty, p\}$, indicating that p apeirogons meet at each vertex. The interior angle θ is then equal to $2\pi / p$. To find the minimum edge length, solve the equation,

$$1 - \tanh\left(\frac{s}{2}\right)\sec\left(\frac{\theta}{2}\right) = \tanh\left(\frac{s}{2}\right)\sec\left(\frac{\pi}{p}\right).$$

For example, if $p = 3$, then,

$$s = 2\tanh^{-1}\left(\frac{1}{2}\right) = \ln 3.$$

Of course, a longer edge length could also be used for this tiling.

There are no apeirogons on the sphere, but there is a degenerate regular tiling of the Euclidean plane with apeirogons with Schläfli symbol $\{\infty, 2\}$. To construct it, divide a line into equal segments which are the edges of the apeirogons. The internal angles are π, and the interiors of the two apeirogons that tile the plane are the two half planes on either side of the line.

Tessellations

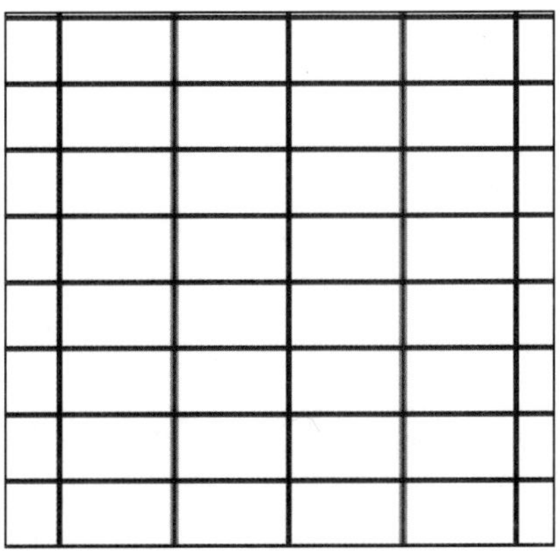

A tessellation with rectangular tiles.

A tessellation is a covering of the plane by shapes, called tiles, so that there are no empty spaces and no overlapped tiles. Tessellations are also called tilings.

Some tessellations involve many types of tiles, but the most interesting tessellations use only one or a few different tiles to fill the plane. In particular, tessellations that use only one kind of tile are called monohedral.

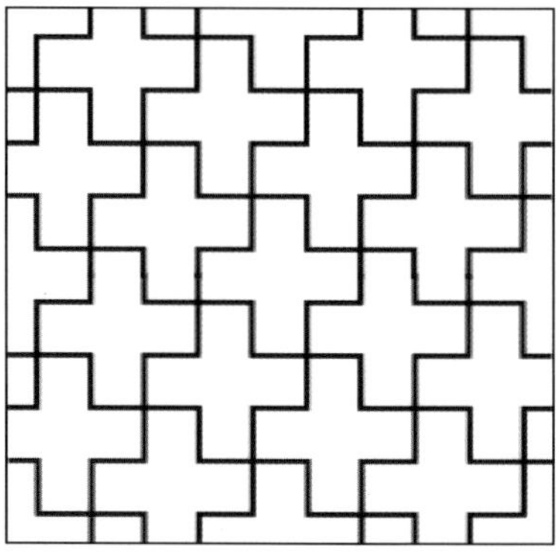

A tessellation with + shaped tiles.

If the tiles are regular polygons (all sides are congruent and all angles are congruent) and vertices only meet vertices we say that the tessellation is a regular tessellation.

An example of tessellations using only a few different tiles are the semi-regular tessellations. These tessellations are made up of two or more types of regular polygons, vertices only meet vertices and the configuration around every vertex is the same.

This Penrose tiling uses two tile shapes:  and .

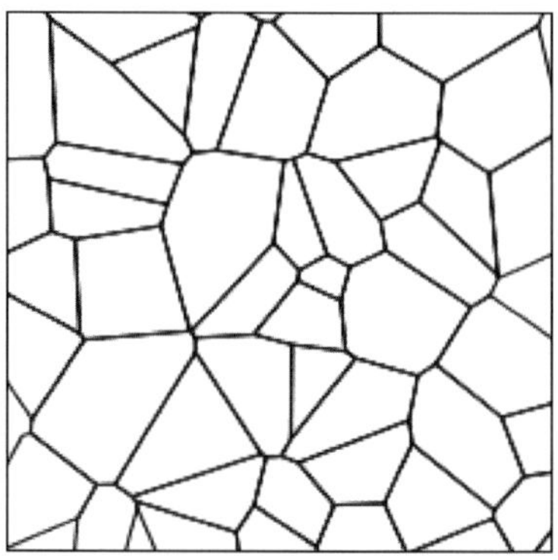

This tiling uses many different polygon tiles.

The rectangle tile tessellation is monohedral, since every tile is the same shape and size. The tessellation by + shapes is monohedral. The Penrose tiling is not monohedral. The tessellations shown here are neither regular nor semi-regular.

Escher's Tessellations

Creating tessellations of the plane by recognizable figures was Escher's first groundbreaking artistic technique. He called it the regular subdivision of the plane, or regelmatige vlakverdeling in his native Dutch. Escher's works feature complex and whimsical figures, mostly animals, which magically interlock to cover the printed page. Creating these mathematically constrained outlines is no easy process, and Escher was unquestionably the master. His technique was to start with a simple tessellation by geometric shapes, and then evolve it into a recognizable figure.

One can divide a plane into geometric shapes according to certain systems; these forms can become infinitely varied and very complex and still meet the requirement of filling the plane with congruent forms in rhythmic repetition without leaving any 'void.' If one then tries to mold the form so as to evoke an association with something familiar to the viewer - an object, an animal, or whatever it might be - it becomes a compelling game in view of the infinite possibilities offered as well as the rigourous restrictions imposed by the rules. It is a painstaking process of groping and fumbling about. Sometimes not a single one of the patterns that appear at the end were consciously sought at the start. It dawns on you slowly. Then comes the moment when you recognize it when you suddenly realize - and here we shift from a general impression to a specific one -: it is a man on a horse! After this, he develops rapidly under the impact of certainty.

Here, Escher lays out his methods for creating tessellations:

- Begin with a tessellation by geometric shapes.
- "Mold the form". Here, he bends and manipulates the straight lines of the geometric scaffolding.
- Look at the resulting pattern and attempt to recognize a figure.
- Push the pattern towards the desired shape.

Not only was this Escher's process for creating tessellations, but also the theme of many of his artworks. Escher wants his audience to see and understand the process behind the patterns, and often presents prints with story like transitions from geometric to recognizable tessellations.

What makes the creation of tessellations a "compelling game" is the interplay between "molding a form" and attempting to recognize a figure in it. We shall see that it is quite difficult to create a tessellation with a preconceived subject or motif, but easy to make interesting looking shapes that, after the fact, can be decorated and enhanced to look like something.

Relevant examples from escher's work:

- Eight Heads,
- Lion and bat patterns on fabric,
- Experiments with Haag's paper,
- Regular divisions of the plane poster,
- Sketches of a horseman,
- Experiments with stamped block tessellations,
- Regular Division of the Plane Drawings,
- Cycle,
- Day and Night,
- Liberation,
- Metamorphosis I,
- Metamorphosis II,
- Metamorphosis III,
- Verbum.

Some Basic Tessellations

Recall that a polygon is a closed plane figure made by joining line segments. More precisely, which polygons can be used as the only tile in a monohedral tessellation of the plane? The most common and simplest tessellation uses a square. Squares easily form horizontal strips:

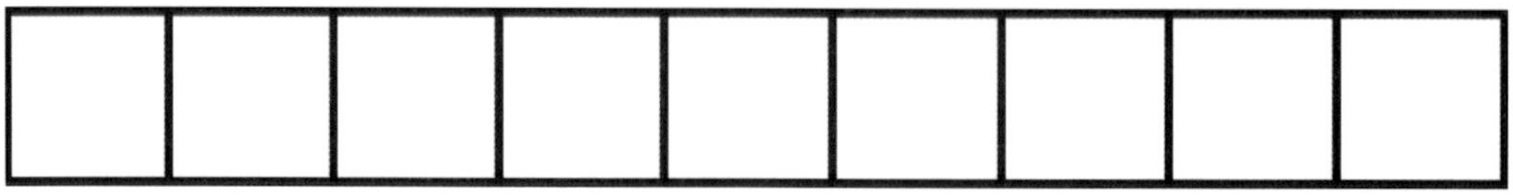

Stacks of these strips cover a rectangular region and the pattern can clearly be extended to cover the entire plane. The same technique works with parallelograms, and so all parallelograms tessellate. Special parallelograms such as rectangles, and rhombuses also tessellate.

Looking for other tessellating polygons is a complex problem, so we will organize the question by the number of sides in the polygon. The simplest polygons have three sides, so we begin with triangles all triangles tessellate. To see this, take an arbitrary triangle and rotate it about the midpoint of one of its sides. The resulting parallelogram tessellates:

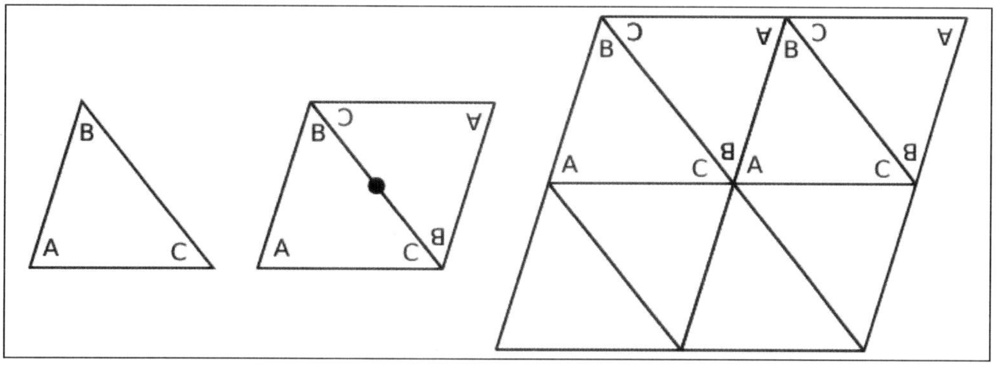

The picture works because all three corners (A, B, and C) of the triangle come together to make a 180° angle - a straight line. This property of triangles will be the foundation of our study of polygon tessellations, so we state it here the sum of angles of any triangle is 180°.

Moving up from triangles, we turn to four sided polygons, the quadrilaterals.

Tessellations by Quadrilaterals

Recall that a quadrilateral is a polygon with four sides the sum of angles in any quadrilateral is 360°

To prove, divide a quadrilateral into two triangles as shown:

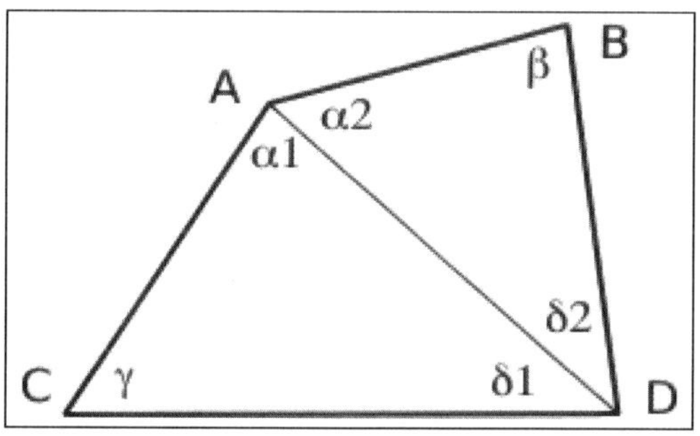

Since the angle sum of any triangle is 180°, and there are two triangles, the angle sum of the quadrilateral is 180° + 180° = 360°. Taking a little more care with the argument, we have:

$$\alpha_1 + \delta_1 + \gamma = 180° \text{ and } \alpha_2 + \delta_2 + \beta = 180°$$

Then,

$$360° = \alpha_1 + \delta_1 + \gamma + \alpha_2 + \delta_2 + \beta = \alpha_1 + \alpha_2 + \beta + \gamma + \delta_1 + \delta_2 = \angle A + \angle B + \angle C + \angle D.$$

The point of all the letters is that the angles of the triangles make the angles of the quadrilateral, which would not work if the quadrilateral was divided as shown on the right.

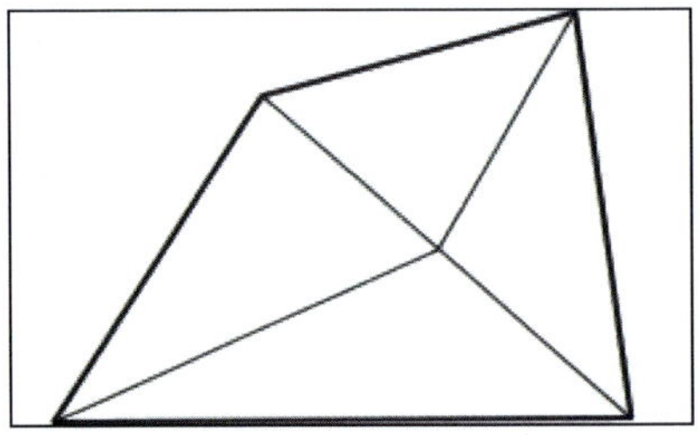

This division into triangles does not calculate the angle sum of the quadrilateral. Begin with an arbitrary quadrilateral ABCD. Rotate by 180° about the midpoint of one of its sides, and then repeat using the midpoints of other sides to build up a tessellation.

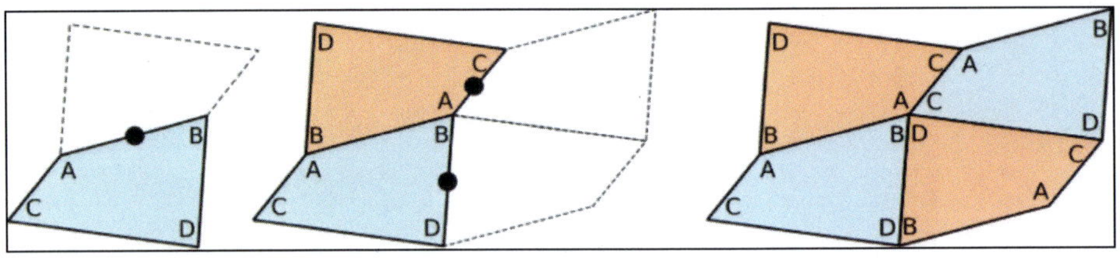

The angles around each vertex are exactly the four angles of the original quadrilateral. Since the angle sum of the quadrilateral is 360°, the angles close up, the pattern has no gaps or overlaps, and the quadrilateral tessellates.

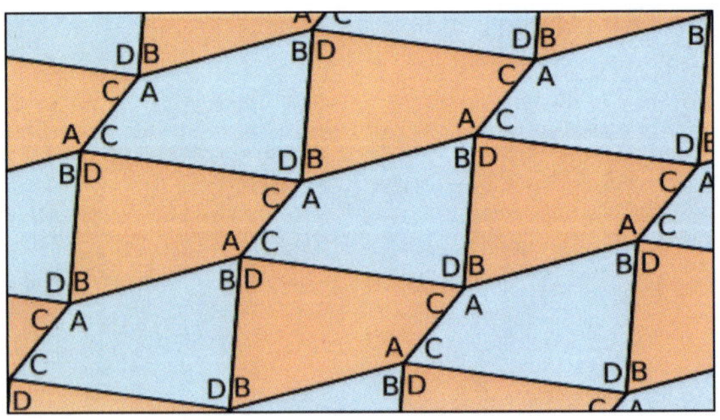

Recall from Fundamental Concepts that a convex shape has no dents. All triangles are convex, but there are non-convex quadrilaterals. The technique for tessellating with quadrilaterals works just as well for non-convex quadrilaterals:

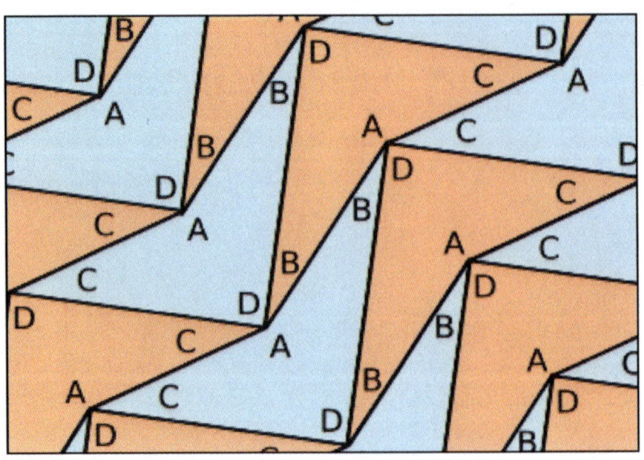

It is worth noting that the general quadrilateral tessellation results in a wallpaper pattern with p2 symmetry group.

Tessellations by Convex Polygons

Every shape of triangle can be used to tessellate the plane. Every shape of quadrilateral can be used to tessellate the plane. In both cases, the angle sum of the shape plays a key role. Since triangles have angle sum 180° and quadrilaterals have angle sum 360°, copies of one tile can fill out the 360° surrounding a vertex of the tessellation.

The next simplest shape after the three and four sided polygon is the five sided polygon: the pentagon. The angle sum of any pentagon is 540°, because we can divide the pentagon into three triangles:

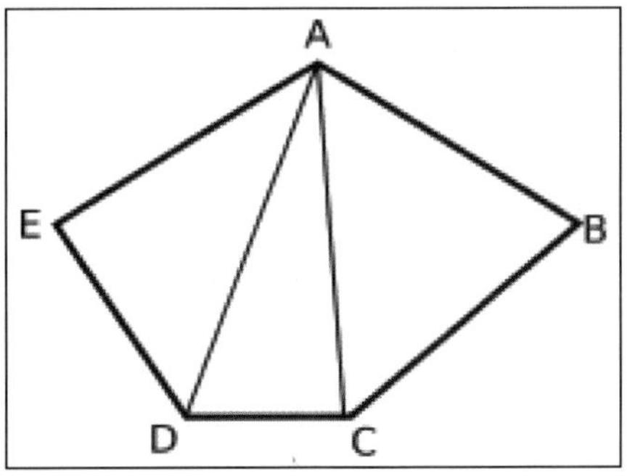

Since each triangle has angle sum 180° the angle sum of the pentagon is 180° + 180° + 180° = 540°.

Rather than repeat the angle sum calculation for every possible number of sides, we look for a pattern. The angle sum of a triangle (3-gon) is 180°, the angle sum of a quadrilateral (4-gon) is 2 × 180°, and the angle sum of a pentagon is 3 × 180°. A general polygon with n sides can be cut into $n-2$ triangles and so we have the sum of the angles of an n-gon is (n − 2) × 180°.

Unlike the triangle and quadrilateral case, the pentagon's angle sum of 540° is not helpful when trying to fit a bunch of pentagons around a vertex. In fact, there are pentagons which do not tessellate the plane.

For example, the regular pentagon has five equal angles summing to 540°, so each angle of the regular pentagon is $\frac{540°}{5} = 108°$. Attempting to fit regular polygons together leads to one of the two pictures below:

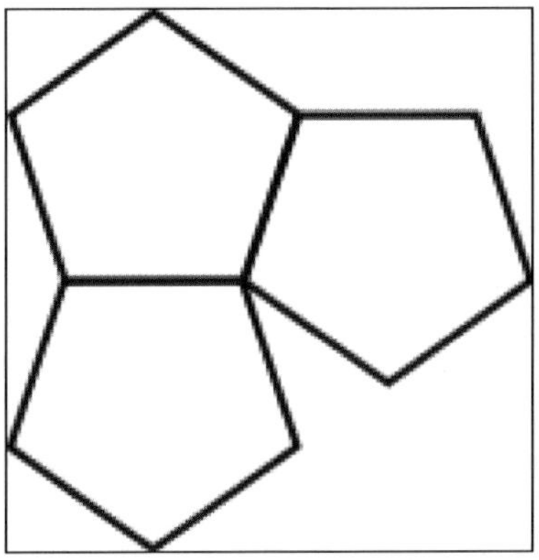

Both situations have wedge shaped gaps that are too narrow to fit another regular pentagon. Thus, not every pentagon tessellates. On the other hand, some pentagons do tessellate, for example this house shaped pentagon: 

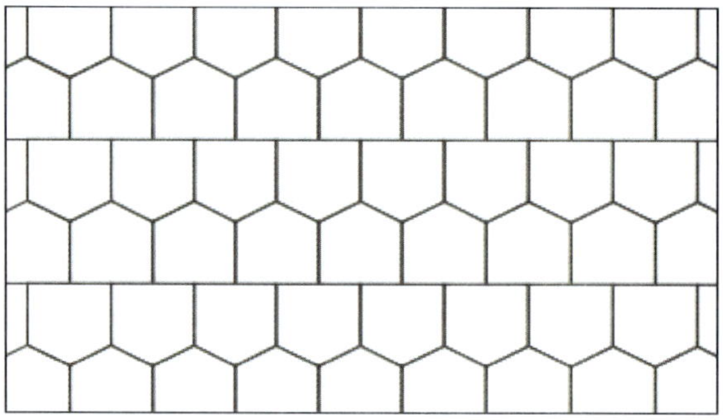

The house pentagon has two right angles. Because those two angles sum to 180° they can fit along a line, and the other three angles sum to 360° (= 540° - 180°) and fit around a vertex.

Thus, some pentagons tessellate and some do not. The situation is the same for hexagons, but for polygons with more than six sides there is the following, no convex polygon with seven or more sides can tessellate.

Nobody has seriously attempted to classify non-convex polygons which tessellate, because the list is quite likely to be too long and messy to describe by hand. However, there has been quite a lot of work towards classifying convex polygons which tessellate.

Reinhardt showed that there are only three types of convex hexagons which tessellate:

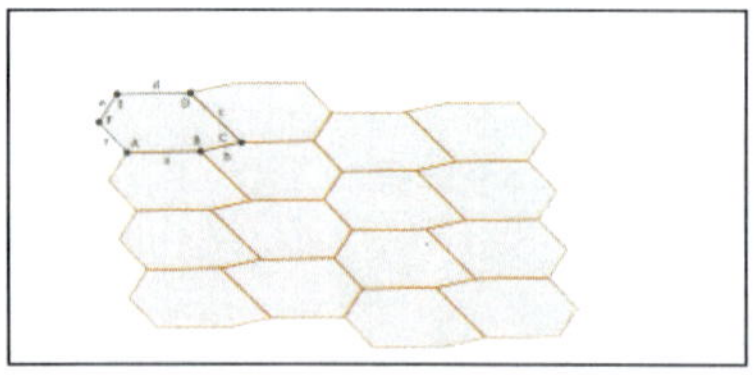	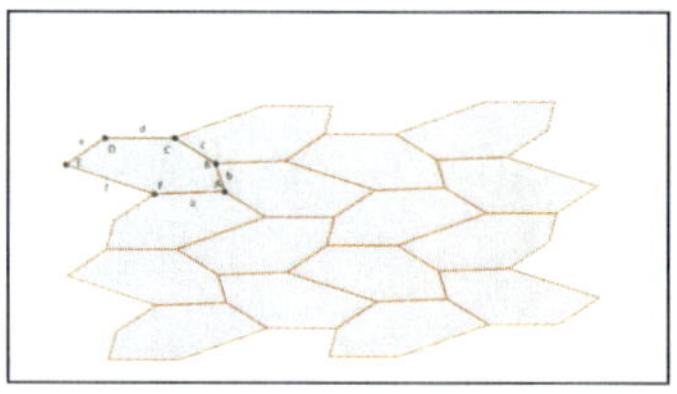	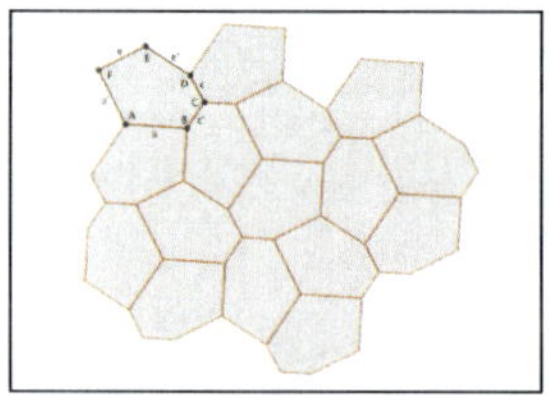
Type 1 B + C + D = 360° A + E + F = 360° a = d	Type 2 A + B + D = 360° C + E + F = 360° a = d c = e	Type 3 A = C = E = 120° a = a' c = c' e = e'

Tessellations by Regular Polygons

Recall that a regular polygon is a polygon whose sides are all the same length and whose angles all have the same measure. A regular n-gon has n equal angles that sum to $(n-2)180°$, so, the corner angle of a regular n-gon is $\frac{(n-2)180°}{n}$.

The table shows the corner angles for the first few regular polygons:

Number of Sides	3	4	5	6	7	8	9	10	11	12
Corner angle	60°	90°	108°	120°	~128.6°	135°	140°	144°	~147.3°	150°

Regular Tessellations

A tessellation using one regular polygon tile, arranged so that edges match up.

Corners of the tiles need to fit together around a point, which means the corner angle of the regular polygon must evenly divide 360°. Since,

$$6 \times 60° = 360°,$$

there is a regular tessellation using six triangles around each vertex. Since,

$$4 \times 90° = 360°,$$

there is a regular tessellation using four squares around each vertex. And since,

$$3 \times 120° = 360°$$

there is a regular tessellation using three hexagons around each vertex. We have already seen that the regular pentagon does not tessellate. A regular polygon with more than six sides has a corner angle larger than 120° (which is 360°/3) and smaller than 180° (which is 360°/2) so it cannot evenly divide 360°. We conclude:

There are three regular tessellations of the plane: by triangles, by squares, by hexagons.

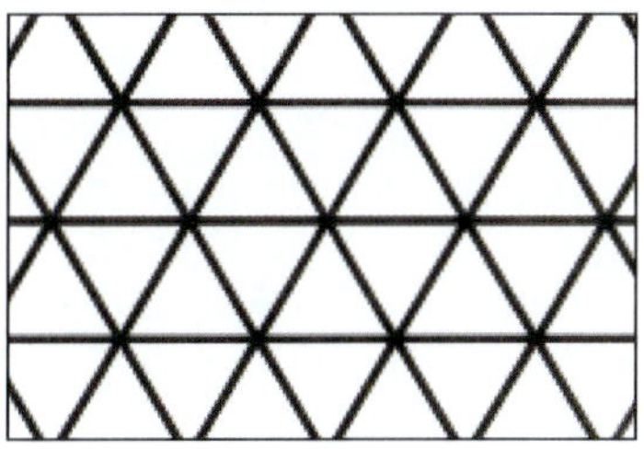

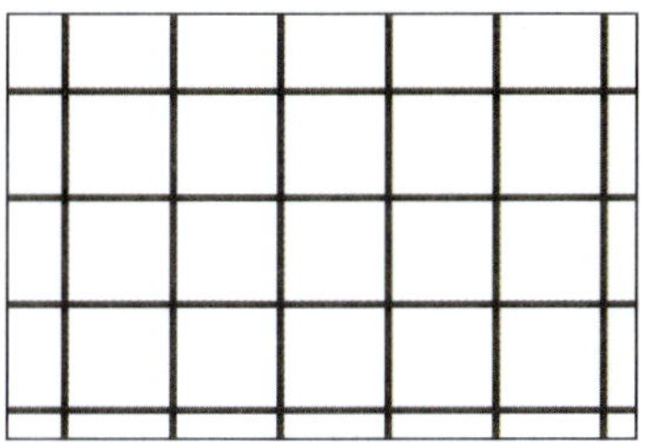

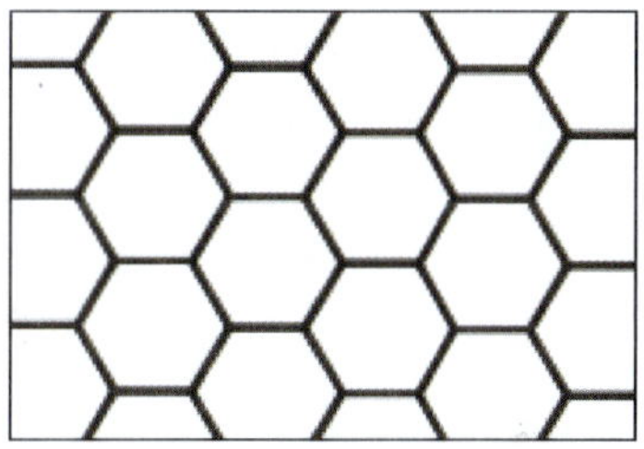

Tessellations using different kinds of regular polygon tiles are fascinating, and lend themselves to puzzles, games, and certainly tile flooring.

Archimedean Tessellations

Floor in Seville.

Side walk in Seville.

An Archimedean tessellation (also known as a semi-regular tessellation) is a tessellation made from more that one type of regular polygon so that the same polygons surround each vertex.

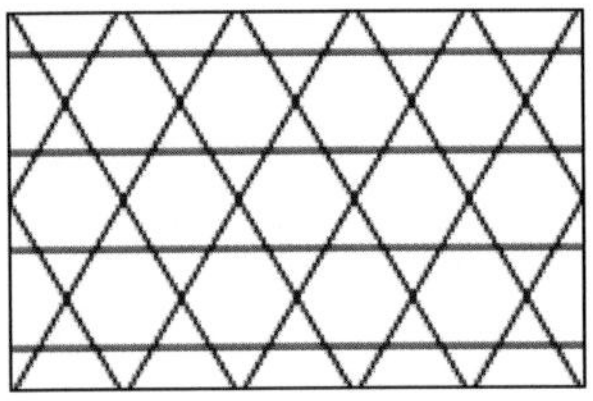

(3,6,3,6)

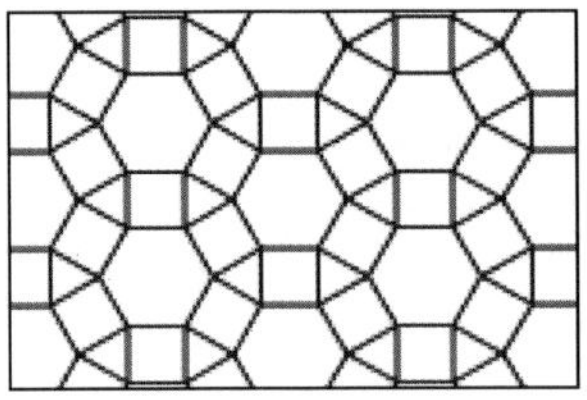

(3,4,6,4)

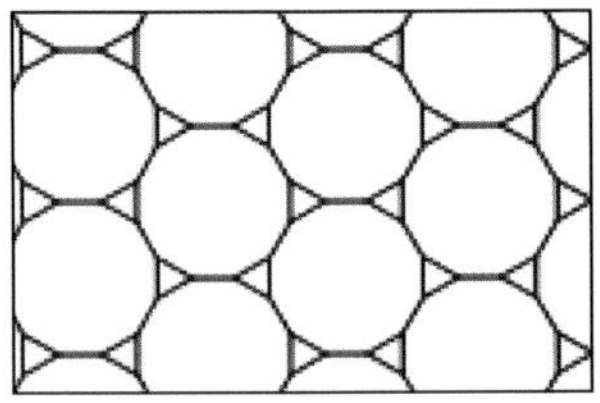

(3,12,12)

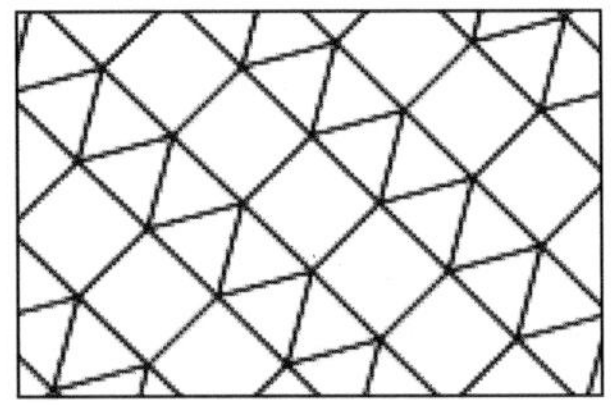

(3,3,3,4,4)

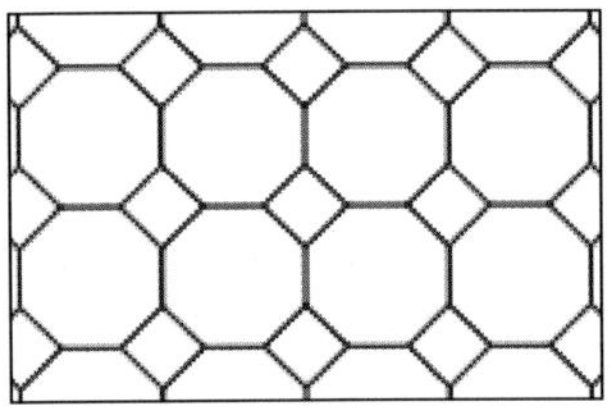

(4,8,8)

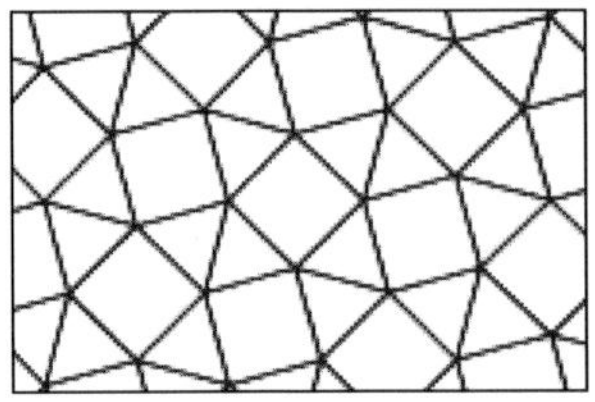

(3,3,4,3,4)

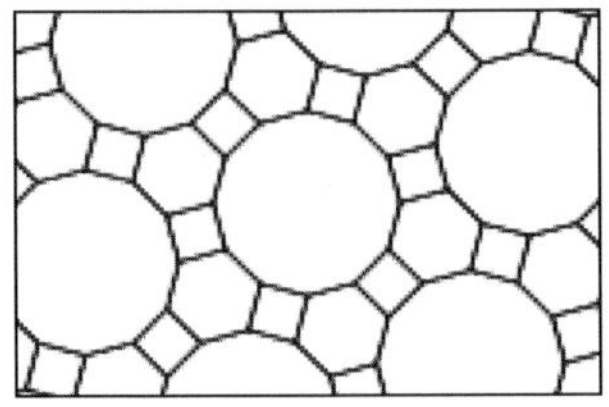

(4,6,12)

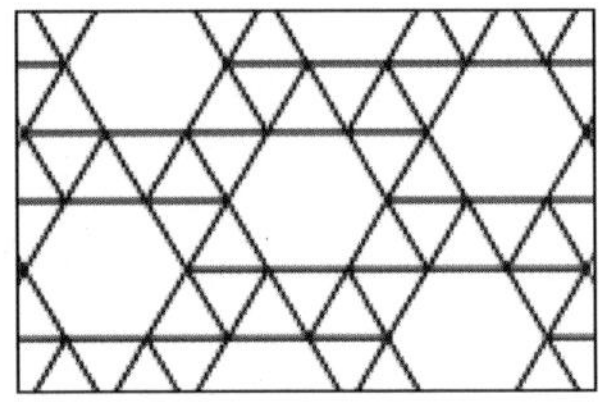

(3,3,3,3,6)

We can use some notation to clarify the requirement that the vertex configuration be the same at every vertex. We can list the types of polygons as they come together at the vertex. For instance in the top row we see on the left a semi-regular tessellation with at every vertex a (3,6,3,6) configuration. We see a 3-gon, a 6-gon, a 3-gon and a 6-gon.

The other tessellations on the top row have a (3,4,6,4), a (3,12,12), and a (3,3,3,4,4) configuration. These configurations are unique up to cyclic reordering (and possibly reversing the order). For example (3,12,12) can also be written as (12,12,3) or (12,3,12). In the bottom row we have (4,8,8), (3,3,4,3,4), (4,6,12) and (3,3,3,3,6) configurations.

Recall that creating a tessellation requires the angle sums of the polygons to add up to 360° around a vertex. The angle of a regular n-gon is $\frac{(n-2)180°}{n}$. If we have a collection of n-gons: $n_1,\ n_2,\ldots,n_k$, then the angles of those polygons need to add up to 360°:

$$\frac{(n_1-2)180°}{n_1}+\frac{(n_2-2)180°}{n_2}+\ldots+\frac{(n_k-2)180°}{n_k}=360°$$

Dividing both sides by 180° gives us the following equation:

$$\frac{n_1-2}{n_1}+\frac{n_2-2}{n_2}+\ldots+\frac{n_k-2}{n_k}=2.$$

For example, suppose that $n_1=3, n_2=3, n_3=3, n_4=4, n_5=4$, then we have: $\frac{1}{3}+\frac{1}{3}+\frac{1}{3}+\frac{2}{4}+\frac{2}{4}=2.$

This means that 3 triangles and 2 squares will give us a vertex type. In this case we can arrange these polygons around the vertex in two different ways: (3,3,3,4,4) and (3,3,4,3,4). Both of these will give rise to a semi-regular tessellation.

Recall that a semi-regular tessellation satisfies the following requirements:

- All the tiles are regular polygons.
- It is a tessellation, hence no gaps or overlaps.
- All the vertices are of the same type.

Gaussian and Mean Curvatures

Let κ_1 and κ_2 be the principal curvatures of a surface patch $\boldsymbol{\sigma}(u, v)$. The Gaussian curvature of σ is,

$$K = \kappa_1 \kappa_2,$$

and its mean curvature is,

$$H = \frac{1}{2}(\kappa_1 + \kappa_2).$$

To compute *K* and *H*, we use the first and second fundamental forms of the surface:

$$Edu^2 + 2F\ dudv + Gdv^2 \text{ and } Ldu^2 + 2M\ dudv + N\ dv^2.$$

Again, we adopt the matrix notation:

$$\mathcal{F}_1 = \begin{pmatrix} E & F \\ F & G \end{pmatrix} \text{ and } \mathcal{F}_2 = \begin{pmatrix} L & M \\ M & N \end{pmatrix}.$$

The principal curvatures are the eigenvalues of $\mathcal{F}_1^{-1}\mathcal{F}_2$. Hence the determinant of this matrix is the product $\kappa_1\kappa_2$, i.e., the Gaussian curvature *K*. So,

$$\begin{aligned} K &= \det\left(\mathcal{F}_1^{-1}\mathcal{F}_2\right) \\ &= \det\left(\mathcal{F}_1\right)^{-1}\det\left(\mathcal{F}_2\right) \\ &= \frac{LN - M^2}{EG - F^2} \end{aligned}$$

The trace of the matrix is the sum of its eigenvalues, thus, twice the mean curvature *H*. After some calculation, we obtain,

$$\begin{aligned} H &= \frac{1}{2} trace\left(\mathcal{F}_1^{-1}\mathcal{F}_2\right) \\ &= \frac{1}{2}\frac{LG - 2MF + NE}{EG - F^2} \end{aligned}$$

An equivalent way to obtain *K* and *H* uses the fact that the principal curvatures are also the roots of,

$$\det\left(\mathcal{F}_2 - \kappa\mathcal{F}_1\right) = 0,$$

which expands into a quadratic equation,

$$\left(EG-F^2\right)\kappa^2-\left(LG-2MF+NE\right)\kappa+LN-M^2=0.$$

The product K and the sum $2H$ of the two roots, can be determined directly from the coefficients. The results are the same as in $=\dfrac{LN-M^2}{EG-F^2}$ and $=\dfrac{1}{2}\dfrac{LG-2MF+NE}{EG-F^2}$.

Conversely, given the Gaussian and mean curvatures K and H, we can easily find the principal curvatures κ_1 and κ_2, which are the roots of,

$$\kappa^2-2H\kappa+K=0,$$

i.e., $H\pm\sqrt{H^2-K}$.

Example: We have considered the surface of revolution,

$$\boldsymbol{\sigma}(u,\ v)=\left(f(u)\cos v,\ f(u)\sin v,\ g(u)\right),$$

where we assume, without loss of generality, that $f>0$ $\dot{f}^2+\dot{g}^2=1$ everywhere. Here a dot denotes d/du. The coefficients of the first and second fundamental forms were determined:

$$E=1,\ F=0,\ G=f^2,\ L=\dot{f}\ddot{g}-\ddot{f}\dot{g},\ M=0,\ N=f\dot{g}.$$

So the Gaussian curvatures is,

$$\begin{aligned}K&=\frac{LN-M^2}{EG-F^2}\\&=\frac{\left(\dot{f}\ddot{g}-\ddot{f}\dot{g}\right)f\dot{g}}{f^2}\\&=\frac{\left(\dot{f}\ddot{g}-\ddot{f}\dot{g}\right)\dot{g}}{f}\end{aligned}$$

Meanwhile, differentiate $\dot{f}^2+\dot{g}^2=1$:

$$\dot{f}\ddot{f}+\dot{g}\ddot{g}=0.$$

Thus,

$$\begin{aligned}\left(\dot{f}\ddot{g}-\ddot{f}\dot{g}\right)\dot{g}&=\dot{f}^2\ddot{f}-\ddot{f}\dot{g}^2\\&=-\ddot{f}\left(\dot{f}^2+\dot{g}^2\right)\\&=-\ddot{f}.\end{aligned}$$

So the Gaussian curvature gets simplified to,

$$K = -\frac{\ddot{f}}{f}.$$

Example: Here we compute the Gaussian and mean curvatures of a Monge patch $z = f(x, y)$. Namely, the patch is described by $\boldsymbol{\sigma}(x, y) = (x, y, f(x, y))$. First, we obtain the first and second derivatives:

$$\boldsymbol{\sigma}_x = (1, 0, f_x), \quad \boldsymbol{\sigma}_y = (0, 1, f_y), \quad \boldsymbol{\sigma}_{xx} = (0, 0, f_{xx}), \quad \boldsymbol{\sigma}_{xy} = (0, 0, f_{xy}), \quad \boldsymbol{\sigma}_{yy} = (0, 0, f_{yy}).$$

Immediately, the coefficients of the first fundamental form are determined,

$$E = 1 + f_x^2, \quad F = f_x f_y, \quad G = 1 + f_y^2.$$

So is the unit normal to the patch:

$$\hat{n} = \frac{\boldsymbol{\sigma}_x \times \boldsymbol{\sigma}_y}{\left\|\boldsymbol{\sigma}_x \times \boldsymbol{\sigma}_y\right\|}$$

$$= \frac{(-f_x, -f_y, 1)}{\sqrt{1 + f_x^2 + f_y^2}}.$$

With the normal $\hat{n}$, we obtain the coefficients of the second fundamental form:

$$L = \boldsymbol{\sigma}_{xx} \cdot \hat{n} = \frac{f_{xx}}{\sqrt{1 + f_x^2 + f_y^2}},$$

$$M = \boldsymbol{\sigma}_{xy} \cdot \hat{n} = \frac{f_{xy}}{\sqrt{1 + f_x^2 + f_y^2}},$$

$$N = \boldsymbol{\sigma}_{yy} \cdot \hat{n} = \frac{f_{yy}}{\sqrt{1 + f_x^2 + f_y^2}}.$$

Plug the expressions for *E, F, G, L, M, N* into $= \frac{LN - M^2}{EG - F^2}$ and $= \frac{1}{2}\frac{LG - 2MF + NE}{EG - F^2}$. A few more steps of symbolic manipulation yield:

$$K = \frac{LN - M^2}{EG - F^2}$$

$$= \frac{f_{xx} f_{yy} - f_{xy}^2}{\left(1 + f_x^2 + f_y^2\right)^2},$$

$$H = \frac{1}{2}\frac{LG - 2MF + NE}{EG - F^2}$$

$$= \frac{f_{xx}\left(1+f_y^2\right)-2f_{xy}f_xf_y+f_{yy}\left(1+f_x^2\right)}{2\left(1+f_x^2+f_y^2\right)^{3/2}}.$$

Classification of Surface Points

The Gaussian curvature is independent of the choice of the unit normal $\hat{n}$. To see why, suppose $\hat{n}$ is changed to $-\hat{n}$ Then the signs of the coefficients of *L, M, N* change, so do the signs of both principal curvatures κ_1 and κ_2, which are the roots of det $(\mathcal{F}_2 - \kappa\mathcal{F}_1)$. Their product $K = \kappa_1\kappa_2$ is unaffected. The mean curvature $H = (\kappa_1+\kappa_2)/2$, nevertheless, has its sign depending on the choice of $\hat{n}$.

The sign of *K* at a point ***p*** on a surface *S* has an important geometric meaning, which is detailed below:

1. $K > 0$ The principal curvatures κ_1 and κ_2 have the same sign. The normal curvature κ in any tangent direction ***t*** is equal to $\kappa_1\cos^2\theta + \kappa_2\sin^2\theta$, where θ is the angle between ***t*** and the principal vector corresponding to κ_1. So κ has the same sign as that of κ_1 and κ_2. The surface is bending away from its tangent plane in all tangent directions at ***p***. The quadratic approximation of the surface near ***p*** is the paraboloid,

$$z = \frac{1}{2}\left(\kappa_1x^2 + \kappa_2y^2\right).$$

We call *p* an elliptic point of the surface. The left figure below plots an elliptic paraboloid $z = x^2 + 2y^2$ with principal curvatures 2 and 4 at the origin.

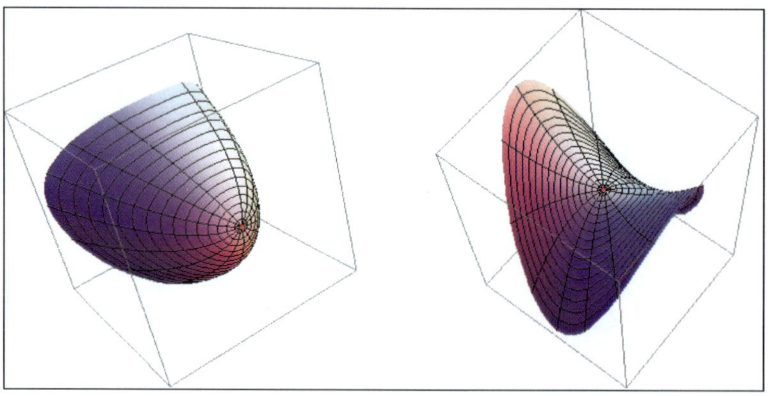

2. $K < 0$ The principal curvatures κ_1 and κ_2 have opposite signs at ***p***. The quadratic approximation of the surface near ***p*** is a hyperboloid. The point is said to be a hyperbolic point of the surface. The right figure above plots a hyperbolic paraboloid $z = x^2 - 2y^2$ with principal curvatures 2 and −4 at the origin.

3. $K = 0$ There are two cases:

- Only one principal curvature, say, κ_1, is zero. In this case, the quadratic approximation is the cylinder $z = \frac{1}{2}\kappa_2y^2$. The point ***p*** is called a parabolic point of the surface.

- Both principal curvatures are zero. The quadratic approximation is the plane $z = 0$. The point $\boldsymbol{p}$ is a planar point of the surface. One cannot determine the shape of the surface near $\boldsymbol{p}$ without examining the third or higher order derivatives. For example, a point in the plane and the origin of a monkey saddle $z = x^3 - 3xy^2$ are both planar points, but they have quite different shapes.

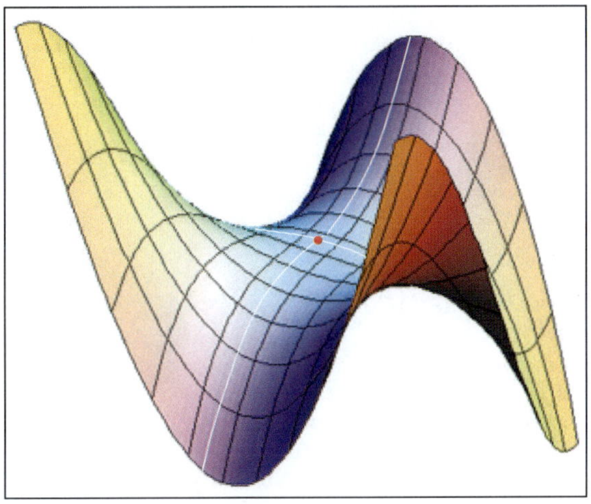

A torus is the surface swept by a circle of radius a originally in the yz-plane and centered on the y-axis at a distance b, $b > a$, from the origin, when the circle revolves about the z-axis. It is easy to derive the following implicit equation for the torus:

$$\left(\sqrt{x^2 + y^2} - b\right)^2 + z^2 = a^2.$$

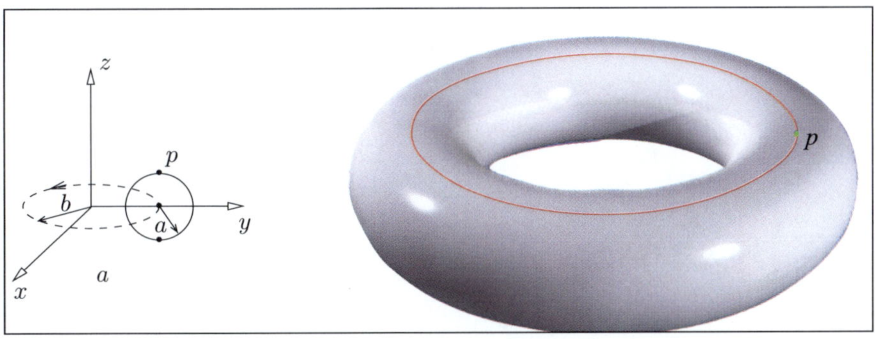

The torus is a good example which has all three types of points. At points on the outer half of the torus, the torus bends away from from its tangent plane; hence $K > 0$. At each point on the inner half, the torus bends toward its tangent plane in the horizontal direction, but away from it in the orthogonal direction; hence $K < 0$. On the two circles, swept respectively by the top and bottom points of the original circle, every point has $K = 0$.

A surface S is flat if its Gaussian curvature is zero everywhere. A plane is flat. Let it be the xy-plane with the parametrization $(x, y, 0)$. We can easily show that the plane has zero Gaussian curvature. A circular cylinder, treated in example of the notes “Surface Curvatures”, has one principal curvature equal to zero and the other equal to the inverse of the radius of its cross section. So a circular cylinder is also flat, even though it is so obviously curved.

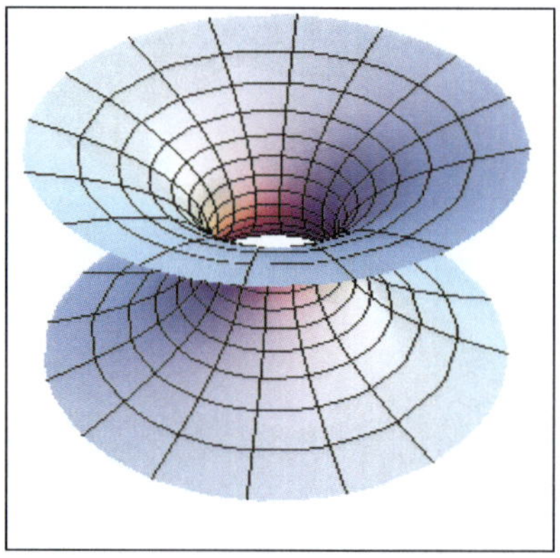

A surface is minimal provided its mean curvature is zero everywhere. Minimal surfaces have Gaussian curvature $K \le 0$. This is because $H = (\kappa_1 + \kappa_2)/2 = 0$ implies $\kappa_1 = -\kappa_2$. The catenoid (cos *u* cosh *v*, sin *u* cosh *v*, *v*) plotted over $[0,\ 2\pi] \times [-2,\ 2]$ to the right is a minimal surface.

The Gauss Map

The standard unit normal $\hat{n}$ to a surface patch σ measures the 'direction' of its tangent plane. The change rate of $\hat{n}$ in a tangent direction, i.e., the normal curvature, indicates the degree of variation of surface geometry in that direction at the point. To make the notion of change of geometry independent of any tangent direction, we can measure by the 'rate of change of $\hat{n}$ per unit area'."

Note that $\hat{n}$ is a point of the unit sphere S^2 centered at the origin. The Gauss map from a surface patch $\boldsymbol{\sigma}(u,\ v): U \to \mathbb{R}^3$ to the unit sphere S^2 sends a point $\boldsymbol{p} = \sigma(u, v)$ to the point $\hat{n}$ (u, v) of S^2. The Gauss map may be a many-to-one mapping since multiple points on the patch can have the same unit normal.

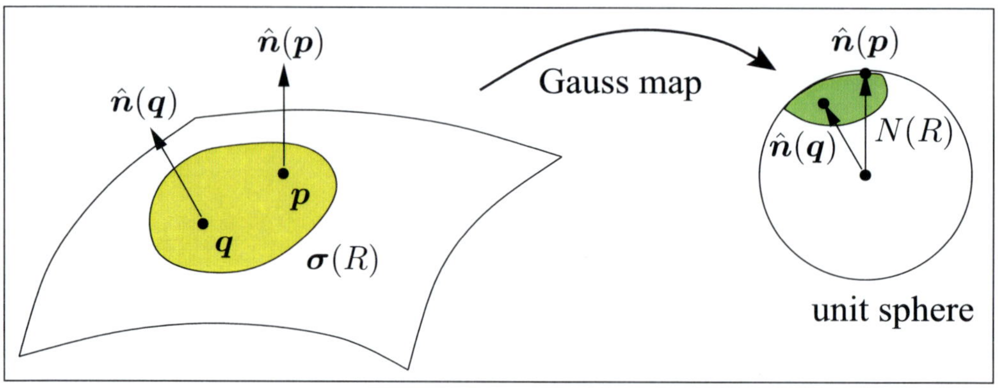

Let $R \subseteq U$ be a region in the patch's domain. The amount by which $\hat{n}$ varies over the corresponding region $\sigma(R)$ on the surface is measured by the area of the image region *N(R)* on the unit sphere. The rate of change of $\hat{n}$ per unit area is the limit of the ratio of the area A_N *(R)* of *N(R)* to the area A_σ*(R)* of the surface region σ*(R)*, as *R* shrinks to a point. To be more precise, we consider *R* to be a closed disk of radius δ centered at *(u, v)* $\in$ *U*. This ratio is,

$$\lim_{\delta \to 0} \frac{\mathcal{A}_N(R)}{\mathcal{A}_\sigma(R)}.$$

It can be shown that the above ratio is the absolute value of the Gaussian curvature at $\boldsymbol{p}$, i.e.,

$$\lim_{\delta \to 0} \frac{A_N(R)}{A_\sigma(R)} = |K|.$$

The integral of the Gaussian curvature K over a surface *S*,

$$\iint_S K dS,$$

is called the total Gaussian curvature of *S*. It is the algebraic area of the image of the region on the unit sphere under the Gauss map. Note the use of the word 'algebraic' since Gaussian curvature can be either positive or negative.

Suppose the patch $S = \sigma(u, v)$ is defined over the domain $[a, b] \times [c, d]$. Then the total Gaussian curvature is computed as,

$$\int_c^d \int_a^b K(u,v)\sqrt{EG - F^2}\, dudv.$$

Example: If the Gaussian curvature *K* of a surface *S* is constant, then the total Gaussian curvature is $K\mathcal{A}(S)$, where $\mathcal{A}(S)$ is the area of the surface. Thus a sphere of radius *r* has total Gaussian curvature $\frac{1}{r^2} \cdot 4\pi r^2$, which is independent of the radius *r*.

Example: Without any computation, we can determine that an ellipsoid also has total curvature 4π. The Gauss map is bijective (one-to-one and onto) since every point on the ellipsoid has a distinct normal. The image region covers the unit sphere. Because the Gaussian curvature is everywhere positive on the ellipsoid, the area of the unit sphere, 4π, is the total Gaussian curvature of the ellipsoid.

Models of the Hyperbolic Plane

Poincarés Disk Model for Hyperbolic Geometry

When we adopt the Hyperbolic Axiomthen there are certain ramifications:

1. The sum of the angles in a triangle is less than two right angles.
2. All similar triangles that are congruent, i.e. AAA is a congruence criterion.
3. There are no lines everywhere equidistant from one another.
4. If three angles of a quadrilateral are right angles, then the fourth angle is less than a right angle.
5. If a line intersects one of two parallel lines, it may not intersect the other.
6. Lines parallel to the same line need not be parallel to one another.
7. Two lines which intersect one another may both be parallel to the same line.

How can we visualize this? Surely it cannot be by just looking at the Euclidean plane in a slightly different way. We need a model with which we could study the hyperbolic plane. If it is to be a Euclidean object that we use to study the hyperbolic plane, $\mathscr{H}^2$, then we must have to make some major changes in our concept of point, line, and distance.

We need a model to see what $\mathscr{H}^2$ looks like. We know that it will not be easy, but we do not want some extremely difficult model to construct.

There are three traditional models for $\mathscr{H}^2$. They are known as the Klein model, the Poincare Disk model, and the Poincarés Half-Plane model. We will start with the Disk model and move to the Half-Plane model later. There are geometric "isomorphisms" between these models, it is just that some properties are easier to see in one model than the other. The two Poincare models tend to give us the opportunity to do computations more easily than the Klein model.

In order to give a model for $\mathscr{H}^2$, we need to decide on a set of points, then determine what lines are and how to measure distance. For Poincarés Disk Model we take the set of points that lie inside the unit circle, i.e., the set,

$$\mathscr{H}^2 = \left\{(x, y) \middle| x^2 + y^2 < 1\right\}.$$

Note that points on the circle itself are not in the hyperbolic plane. However they do play an important part in determining our model. Euclidean points on the circle itself are called ideal points, omega points, vanishing points, or points at infinity.

Poincarés himself thought of this set as the set of all complex numbers with length less than 1,

$$\mathscr{H}^2 = \left\{z \in \mathbb{C} \middle| \, \|z\| < 1\right\}.$$

We will see why this is important when we study the Poincarés half plane model.

A unit circle is any circle in the Euclidean plane is a circle with radius one.

Given a unit circle Γ in the Euclidean plane, points of the hyperbolic plane are the points in the interior of Γ. Points on this unit circle are called omega points (Ω) of the hyperbolic plane.

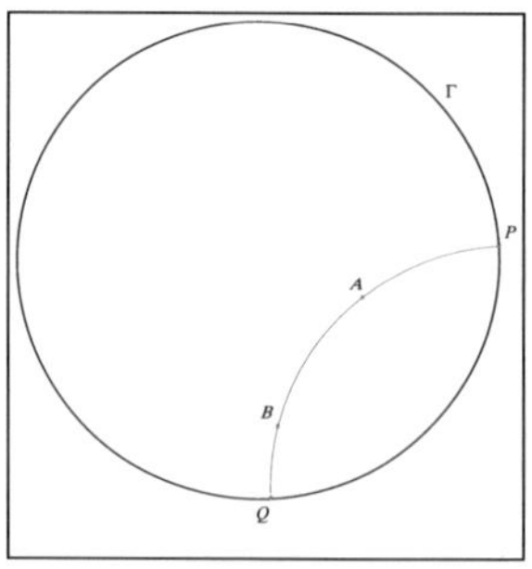

Poincaré line.

If we take Γ to be the unit circle centered at the origin, then we would think of the hyperbolic plane as $\mathscr{H}^2 = \left\{(x, y) \middle| x^2 + y^2 < 1\right\}$ and the omega points are the points $\Omega = \left\{(x, y) \middle| x^2 + y^2 = 1\right\}$. The points in the Euclidean plane satisfying $\left\{(x, y) \middle| x^2 + y^2 > 1\right\}$ are called ultraideal points.

We now have what our points will be. We see that we are going to have to modify our concept of line in order to have the Hyperbolic Axiom to hold.

Definition: Given a unit circle Γ in the Euclidean plane, lines of the hyperbolic plane are arcs of circles drawn orthogonal to Γ and located in the interior of Γ.

Construction of Lines

This sounds nice, but how do you draw them?

1. Start with a circle Γ centered at O and consider the ray $\overrightarrow{OA}$, where A lies on the circle, Γ.
2. Construct the line perpendicular to $\overrightarrow{OA}$ at A.
3. Choose a point P on this perpendicular line for the center of the second circle and make P A the radius of a circle centered at P.
4. Let B denote the second point of intersection with circle Γ. Then the arc AB represents a line in this model.

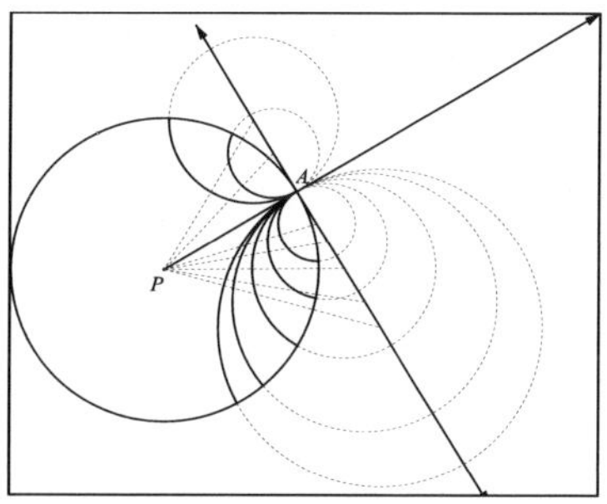

Poincare lines through A.

Now, how do you construct these lines in more general circumstances? There are three cases we need to consider.

Case I: $A,\ B \in \Gamma$

Case II: $A \in \Gamma$ and B lies inside Γ

Case III: A and B both lie inside Γ

Case I: Construct rays $\overrightarrow{PA}$ and $\overrightarrow{PB}$ where P is the center of the circle Γ. Construct the lines perpendicular to $\overrightarrow{PA}$ and $\overrightarrow{PB}$ at A and B respectively. Let Q be the point of intersection of those two lines. The circle Ω centered at Q with radius QA intersects Γ at A and B. The line between A and B is the arc of Ω that lies inside Γ.

Note that this arc is clearly orthogonal to Γ by its construction.

Case II: Construct rays $\overrightarrow{PA}$ and $\overrightarrow{PB}$ where P is the center of the circle Γ. Construct the line perpendicular to $\overrightarrow{PA}$ at A. Draw segment AB and construct its perpendicular bisector. Let Q be the point of intersection of this line and the tangent line to Γ at A. The circle Ω centered at Q with radius QA contains A and B. The line containing A and B is the arc of Ω that lies inside Γ.

This arc, as constructed is orthogonal to Γ at A. We want to see that it is orthogonal at the other point of intersection with the circle. Let that point of intersection be X. Then, $X \in \Gamma$ means that PA $\cong$ P X. Since X lies on our second circle it follows that $QX \cong QA$. Since $PQ \cong PQ$, we have that $\Delta PAQ \cong \Delta PXQ$, which means that $\angle PXQ$ is a right angle, as we wanted to show.

Case III: Construct the ray $\overrightarrow{PA}$ and then construct the line perpendicular to $\overrightarrow{PA}$ at A. This intersects Γ in points X and Y. Construct the tangents to Γ at X and at Y. These tangent lines intersect at a point C. The circle Ω centered at Q is the circle passing through A, B, and C. The line containing A and B is the arc of Ω that lies inside Γ.

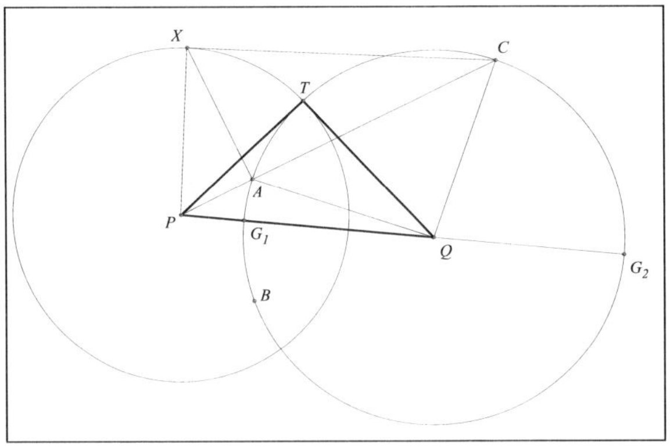

Poincare line in Case III.

From our construction, we have that $\Delta PXC \sim PAX$ and it follows that $|PA||PC| = |PX|^2 = r^2$. Now, Q lies on the perpendicular bisectors of AC and AB as Ω is the circumcircle for ΔABC. There is a point T on the circle Ω so that the tangent line to Ω at T passes through P.

Construct the line through P and Q which intersects the circle in two points G_1 and G_2 so that G_1 lies between P and Q. Now,

$$\begin{aligned}
|PT|^2 &= |PQ|^2 - |QT|^2 \\
&= (|PQ| - |QT|)(|PQ| + |QT|) \\
&= (|PQ| - |QG_1|)(|PQ| + |QG_2|) \\
&= |PG_1||PG_2| \\
&= |PA||PC| = r^2
\end{aligned}$$

Therefore, T lies on the circle Γ and Γ and Ω are orthogonal at that point. A similar argument shows that they are orthogonal at the other point of intersection.

Distance

Now, this area inside the unit circle must represent the infinite hyperbolic plane. This means that our standard distance formula will not work. We introduce a distance metric by,

$$d\rho = \frac{2dr}{1-r^2}$$

where ρ represents the hyperbolic distance and r is the Euclidean distance from the center of the circle. Note that $d\rho \to \infty$ as $r \to 1$. This means that lines are going to have infinite extent.

The relationship between the Euclidean distance of a point from the center of the circle and the hyperbolic distance is:

$$\rho = \int_0^r \frac{2du}{1-u^2} = \log\left(\frac{1+r}{1-r}\right) = 2\tanh^{-1} r,$$

$$\text{or } r = \tanh\frac{\rho}{2}.$$

Now, for those of you who don't remember ever having seen this function tanh(*x*), we give a little review. The hyperbolic trigonometric functions cosh(*x*) and sinh(*x*) are defined by:

$$\sinh(x) = \frac{e^x - e^{-x}}{2}$$

$$\cosh(x) = \frac{e^x + e^{-x}}{2}$$

and

$$\tanh(x) = \frac{\sinh(x)}{\cosh(x)} = \frac{e^x - e^{-x}}{e^x + e^{-x}} = \frac{e^{2x} - 1}{e^{2x} + 1}.$$

Now, we can use this to define the distance between two points on a Poincarés line. Given two hyperbolic points *A* and *B*, let the Poincare line intersect the circle in the omega points *P* and *Q*. Define,

$$(AB, PQ) = \frac{AP / AQ}{BP / BQ} = \frac{AP \cdot BQ}{AQ \cdot BP},$$

to be the cross ratio of *A* and *B* with respect to *P* and *Q*, where *AP* denotes the the Euclidean ar-clength. Define the hyperbolic distance from *A* to *B* to be,

$$d(A,\ B) = \log|AB,\ P\ Q|.$$

Theorem: If a point *A* in the interior of Γ is located at a Euclidean distance $r < 1$ from the center *O*, its hyperbolic distance from the center is given by,

$$d(A, O) = \log\frac{1+r}{1-r}.$$

Lemma: The hyperbolic distance from any point in the interior of Γ to the circle itself is infinite.

Parallel Lines

It is easy to see that the Hyperbolic Axiom works in this model. Given a line $\overleftrightarrow{AB}$ and a point $D \notin \overleftrightarrow{AB}$ then we can draw at least two lines through D that do not intersect $\overleftrightarrow{AB}$.

Call these two lines through D lines ℓ_1 and ℓ_2. Notice now how two of our previous results do not hold, as we remarked earlier. We have that $\overleftrightarrow{AB}$ and ℓ_1 and $\overleftrightarrow{AB}$ and ℓ_2 are parallel, but ℓ_1 and ℓ_2 are not parallel. Note also that ℓ_2 intersects one of a pair of parallel lines (ℓ_1), but does not intersect the other parallel line ($\overleftrightarrow{AB}$).

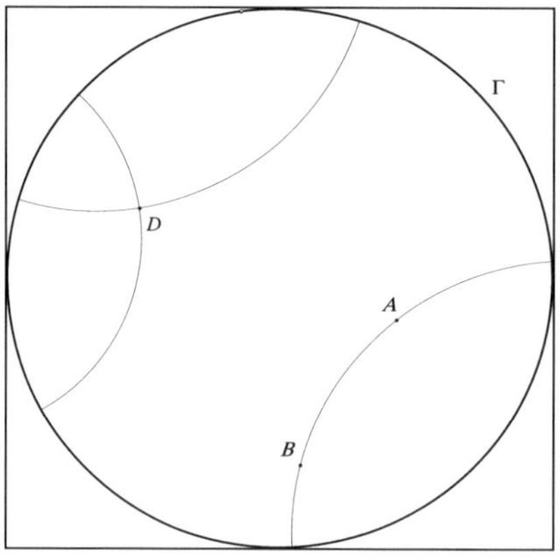

Multiple parallels through *A*.

As we now know, the hyperbolic plane has two types of parallel lines. The definition that we will give here will depend explicitly on the model that we have chosen. Consider the hyperbolic line $\overleftrightarrow{AB}$ which intersects the circle Σ in the ideal points Λ and Ω. Take a point D/∈ $\overleftrightarrow{AB}$. Construct the line through Λ and D. Since this line does not intersect the line $\overleftrightarrow{AB}$ inside the circle, these two hyperbolic lines are parallel. However, they seem to be approaching one another as we go "to infinity". Since there are two "ends" of the Poincare line $\overset{\frown}{AB}$, there are two of these lines. The line $\overset{\frown}{AB}$ and $\overset{\frown}{D\Lambda}$ are horoparallel. The defining property is as follows.

Definition: Let $P \in \overset{\frown}{AB}$. Consider the collection of lines $\overset{\frown}{DP}$ as P goes to Ω or Λ. The first line through D in this collection that does not intersect $\overset{\frown}{AB}$ in $\mathscr{H}^2$ is the horoparallel line to $\overset{\frown}{AB}$ in that direction.

Drop a perpendicular from D to $\overset{\frown}{AB}$ and label this point of intersection M. Angles $\angle\Lambda DM$ and $\angle\Omega DM$ are called angles of parallelism.

Theorem: The angles of parallelism associated with a given line and point are congruent.

Proof: Assume not, i.e., assume $\angle\Lambda DM \neq \angle\Omega DM$. Then one angle is greater than the other. Without loss of generality, we may assume that $\angle\Lambda DM < \angle\Omega DM$. Then there is a point E in the interior of $\angle\Omega DM$ such that $\angle\Lambda DM = \angle EDM$. The line $\overset{\frown}{ED}$ must intersect $\overset{\frown}{AB}$ since $\overset{\frown}{D\Omega}$ is the limiting parallel line to $\overset{\frown}{AB}$ in that direction. Let the point of intersection be F. Choose G on $\overset{\frown}{AB}$ on the opposite side of $\overset{\frown}{DM}$ from F so that $FM = GM$. Then $\Delta GMD \cong \Delta FMD$. This implies that $\angle GDM = \angle FDM = \angle\Lambda DM$. This means that $\overset{\frown}{D\Omega}$ intersects $\overset{\frown}{AB}$ at G. This contradicts the condition that $\overset{\frown}{D\Omega}$ is limiting parallel to $\overset{\frown}{AB}$. Thus, the angles of parallelism are congruent.

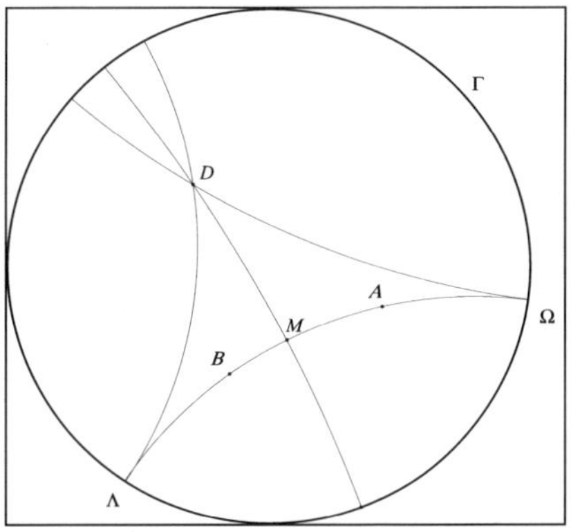

Limiting Parallel Poincare Lines.

Theorem: The angles of parallelism associated with a given line and point are acute.

Proof: Assume not, i.e., assume that $\angle MD\Omega > 90°$. Then there is a point E interior to $\angle MD\Omega$ so that $\angle MDE = 90°$. Then, since $\overset{\frown}{DE}$ and $\overset{\frown}{AB}$ are perpendicular to the same line, they are parallel. Thus, $\overset{\frown}{DE}$ does not intersect $\overset{\frown}{AB}$ which contradicts the condition that $D\Omega$ is the limiting parallel line.

If the angle of parallelism is 90° then we can show that we have Euclidean geometry. Thus, in $\mathscr{H}^2$ the angle of parallelism is acute.

Theorem (Lobachevskii's Theorem): Given a point P at a hyperbolic distance ρ from a hyperbolic line $\overset{\frown}{AB}$ (i.e., $d(P, M) = \rho$), the angle of parallelism, θ, associated with the line and the point satisfies,

$$e^{-\rho} = \tan\left(\frac{\theta}{2}\right).$$

Note then that,

$$\lim_{\rho \to 0} \theta = \frac{\pi}{2} \text{ and } \lim_{\rho \to 0} \theta = 0.$$

Proof: The proof of this is interesting in that we play one geometry against the other in order to arrive at our conclusion.

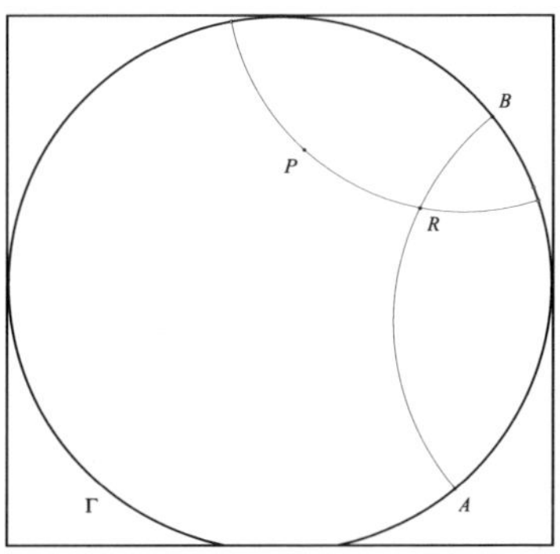

Lobachevskii's Theorem.

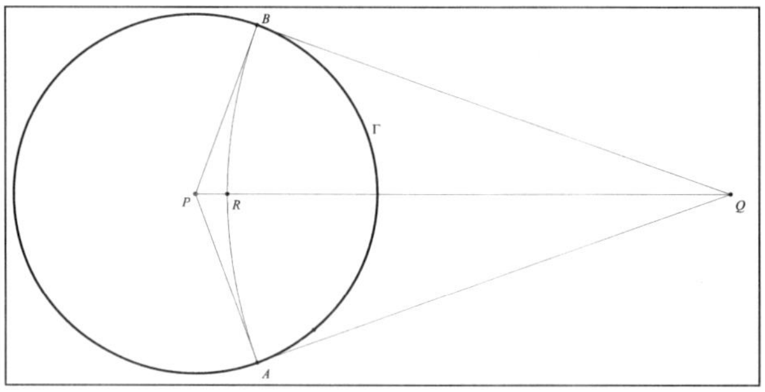

After the first translation.

We are given a line $\widehat{AB}$ and a point P not on the line. Construct the line through P which is perpendicular to $\widehat{AB}$. Call the point of intersection R as in figure. Then we have that $\rho = d(P, R)$. We can translate P to the center of the unit circle and translate our line to a line so that our line perpendicular to AB is a radius of Γ as we have done in figure. Construct the radii from P to the ideal points A and B and construct the lines tangent to Γ at these points. These tangent lines intersect at a point Q which lies on $\overrightarrow{PR}$. Now, since we have moved our problem to the center of the circle, we can use our previous result to see that if r is the Euclidean distance from P to R, then we have,

$$\rho = \log\frac{1+r}{1-r},$$

or rewriting this we have,

$$e^{\rho} = \frac{1+r}{1-r} \text{ or } e^{-\rho} = \frac{1-r}{1+r}.$$

Now, we are talking about Euclidean distances (with r) and using our Euclidean right triangles with radius 1 we have that:

$$r = QP - QR = QP - QA = \sec\ \angle QPA - \tan \angle QPA = \sec\theta - \tan\theta = \frac{1-\sin\theta}{\cos\theta}$$

Now, algebra leads us to:

$$\begin{aligned} e^{-\rho} &= \frac{1-r}{1+r} \\ &= \frac{\cos\theta + \sin\theta - 1}{\cos\theta - \sin\theta + 1} \\ &= \frac{\cos\theta + \ \sin\ \theta - 1\cos\theta + \sin\theta + 1}{\cos\theta - \ \sin\ \theta + 1\cos\theta + \sin\theta + 1} \\ &= \frac{\cos^2\theta + 2\cos\theta\sin\theta + \sin^2\theta - 1}{\cos^2\theta + 2\cos\theta - \sin^2\theta + 1} \\ &= \frac{2\sin\theta\cos\theta}{2\cos^2\theta + 2\cos\theta} = \frac{\sin\theta}{1+\cos\theta} \end{aligned}$$

$$= \frac{2\sin\left(\frac{\theta}{2}\right)\cos\left(\frac{\theta}{2}\right)}{\left(2\cos^2\left(\frac{\theta}{2}\right)-1\right)+1}$$

$$= \tan\left(\frac{\theta}{2}\right)$$

Hyperbolic Circles

Now, if we have a center of a circle that is not at the center P of the unit circle Σ, we know that the hyperbolic distance in one direction looks skewed with respect to the Euclidean distance. That would lead us to expect that a circle in this model might take on an elliptic or oval shape. We will prove later that this is not the case. In fact, hyperbolic circles embedded in Euclidean space retain their circular appearance — their centers are offset.

Theorem: Given a hyperbolic circle with radius R, the circumference + of the circle is given by $C = 2\pi \sinh(R)$.

Common Figures in the Disk Model

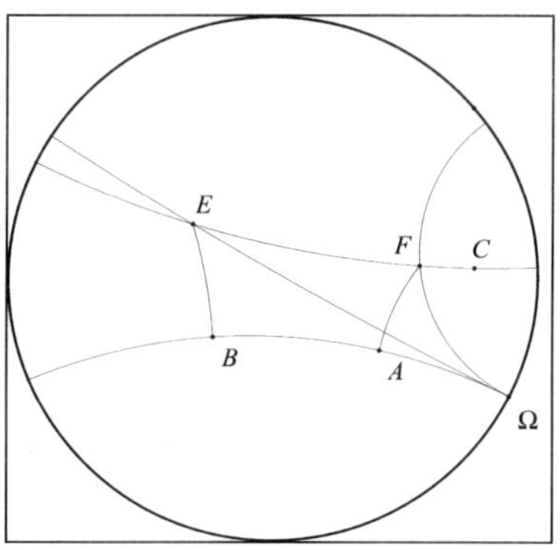

Saccheri quadrilateral in the Poincare Disk.

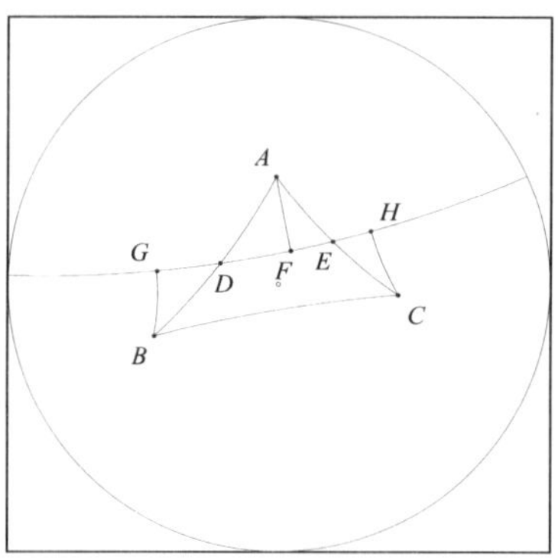

Acute Triangle.

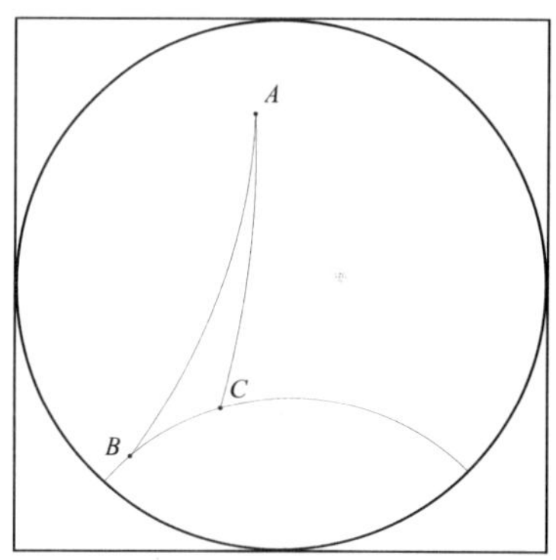

Obtuse Triangle.

Beltrami-Klein Model of Hyperbolic Geometry

In geometry, the Beltrami–Klein model, also called the projective model, Klein disk model, and the Cayley–Klein model, is a model of hyperbolic geometry in which points are represented by the points in the interior of the unit disk (or n-dimensional unit ball) and lines are represented by the chords, straight line segments with ideal endpoints on the boundary sphere.

The Beltrami–Klein model is named after the Italian geometer Eugenio Beltrami and the German Felix Klein while "Cayley" in Cayley–Klein model refers to the English geometer Arthur Cayley. The Beltrami–Klein model is analogous to the gnomonic projection of spherical geometry, in that geodesics (great circles in spherical geometry) are mapped to straight lines.

This model is not conformal, meaning that angles and circles are distorted, whereas the Poincaré disk model preserves these. In this model, lines and segments are straight Euclidean segments, whereas in the Poincaré disk model, lines are arcs that meet the boundary orthogonally.

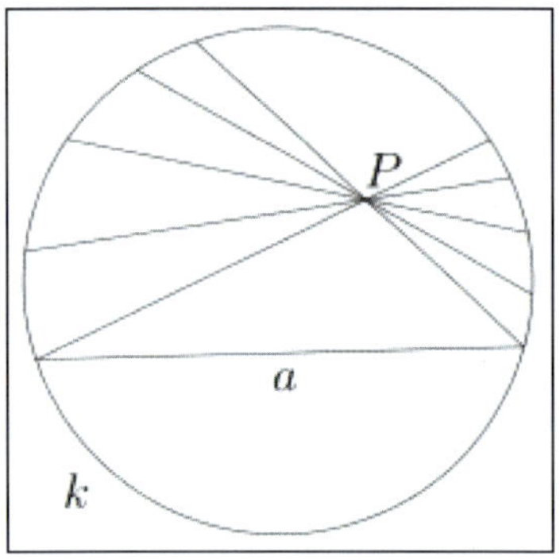

Many hyperbolic lines through point P not intersecting line a in the Beltrami Klein model.

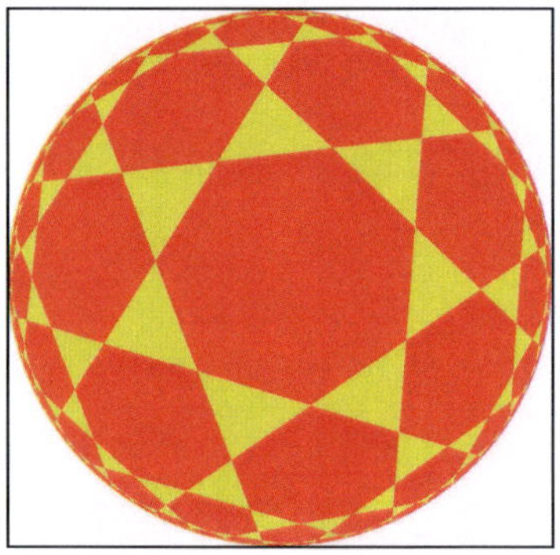

A hyperbolic triheptagonal tiling in a Beltrami–Klein model projection.

Distance Formula

The distance function for the Beltrami–Klein model is a Cayley–Klein metric. Given two distinct points p and q in the open unit ball, the unique straight line connecting them intersects the boundary at two ideal points, a and b, label them so that the points are, in order, a, p, q, b and $|aq| > |ap|$ and $|pb| > |qb|$.

The hyperbolic distance between p and q is then: $d(p,q) = \frac{1}{2}\log\frac{|aq||pb|}{|ap||qb|}$.

The vertical bars indicate Euclidean distances between the points between them in the model, log is the natural logarithm and the factor of one half is needed to give the model the standard curvature of −1.

When one of the points is the origin and Euclidean distance between the points is r then the hyperbolic distance is:

$\frac{1}{2}\ln\left(\frac{1+r}{1-r}\right) = \operatorname{artanh} r$ Where artanh is the inverse hyperbolic function or tangens hyperbolico.

The Klein Disk Model

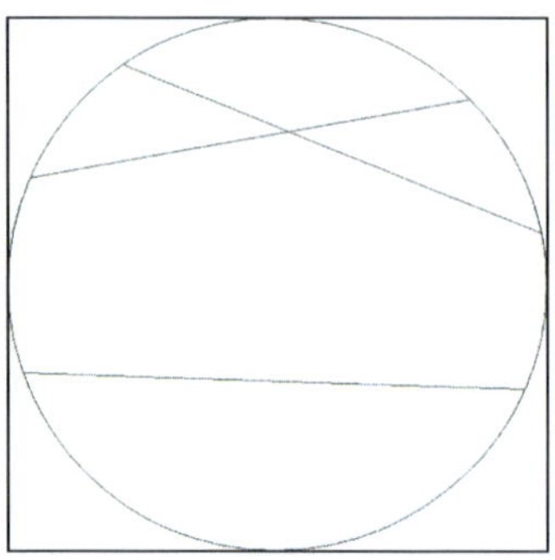

Lines in the projective model of the hyperbolic plane.

In two dimensions the Beltrami–Klein model is called the Klein disk model. It is a disk and the inside of the disk is a model of the entire hyperbolic plane. Lines in this model are represented by chords of the boundary circle (also called the *absolute*). The points on the boundary circle are called ideal points; although well defined, they do not belong to the hyperbolic plane. Neither do points outside the disk, which are sometimes called ultra ideal points.

The model is not conformal, meaning that angles are distorted, and circles on the hyperbolic plane are in general not circular in the model. Only circles that have their centre at the centre of the boundary circle are not distorted. All other circles are distorted, as are horocycles and hypercycles.

Properties

Chords that meet on the boundary circle are limiting parallel lines. Two chords are perpendicular if, when extended outside the disk, each goes through the pole of the other. (The pole of a chord is an ultra ideal point: the point outside the disk where the tangents to the disk at the endpoints of the chord meet.) Chords that go through the centre of the disk have their pole at infinity, orthogonal to the direction of the chord (this implies that right angles on diameters are not distorted).

Compass and Straightedge Constructions

Here is how one can use compass and straightedge constructions in the model to achieve the effect of the basic constructions in the hyperbolic plane.

- The pole of a line. While the pole is not a point in the hyperbolic plane (it is an ultra ideal point) most constructions will use the pole of a line in one or more ways.
- For a line: construct the tangents to the boundary circle through the ideal (end) points of the line. the point where these tangents intersect is the pole.
- For diameters of the disk: the pole is at infinity perpendicular to the diameter.
- To construct a perpendicular to a given line through a given point draw the ray from the pole of the line through the given point. The part of the ray that is inside the disk is the perpendicular.
- When the line is a diameter of the disk then the perpendicular is the chord that is (Euclidean) perpendicular to that diameter and going through the given point.
- To find the midpoint of given segment AB: Draw the lines through A and B that are perpendicular to AB. Draw the lines connecting the ideal points of these lines, two of these lines will intersect the segment AB and will do this at the same point. This point is the (hyperbolic) midpoint of AB.
- To bisect a given angle $\angle BAC$: Draw the rays AB and AC. Draw tangents to the circle where the rays intersect the boundary circle. Draw a line from A to the point where the tangents intersect. The part of this line between A and the boundary circle is the bisector.
- The common perpendicular of two lines is the chord that when extended goes through both poles of the chords.

- When one of the chords is a diameter of the boundary circle then the common perpendicular is the chord that is perpendicular to the diameter and that when lengthened goes through the pole of the other chord.
- To reflect a point P in a line l: From a point R on the line l draw the ray through P. Let X be the idealpoint where the ray intersects the absolute. Draw the ray from the pole of line l through X, let Y be the other intersection point with the absolute. Draw the segment RY. The reflection of point P is the point where the ray from the pole of line l through P intersects RY.

Circles, Hypercycles and Horocycles

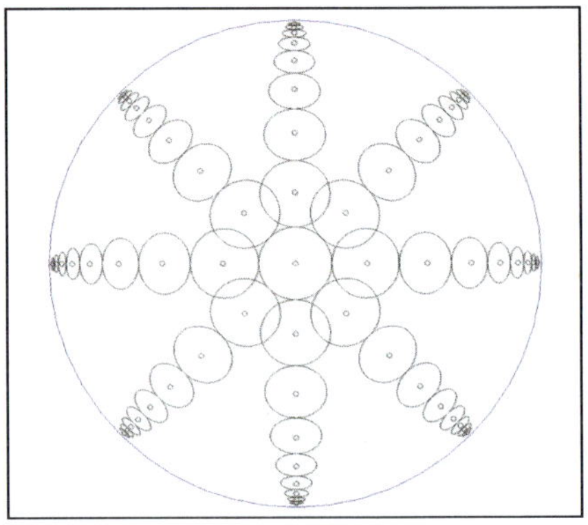

Circles in the Klein disk model.

While lines in the hyperbolic plane are easy to draw in the Klein disk model, it is not the same with circles, hypercycles and horocycles.

Circles (the set of all points in a plane that are at a given distance from a given point, its center) in the model become ellipses increasingly flattened as they are nearer to the edge. Also angles in the Klein disk model are deformed. For constructions in the hyperbolic plane that contain circles, hypercycles, horocycles or non right angles it is better to use the Poincaré disk model or the Poincaré half-plane model.

Relation to the Poincaré Disk Model

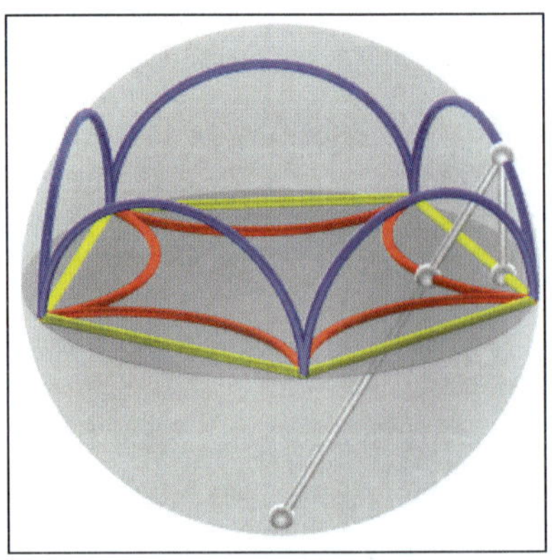

Combined projections from the Klein disk model (yellow) to the Poincaré disk model (red) via the hemisphere model (blue).

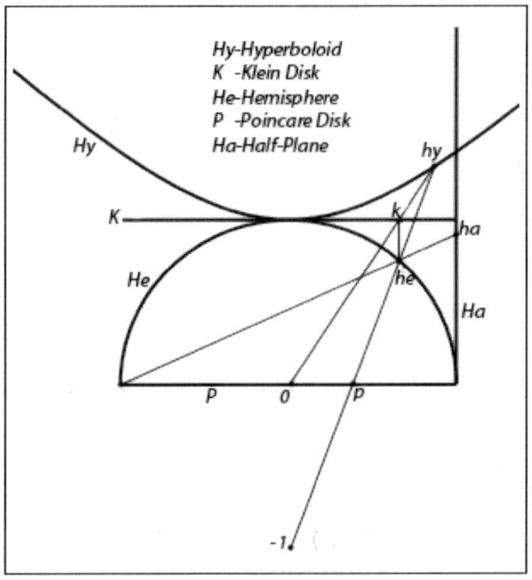

The Beltrami–Klein model (K in the picture) is orthographic projection from hemispherical model. and a gnomonic projection of the hyperboloid model (Hy) with as center the center of the hyperboloid (O).

Both the Poincaré disk model and the Klein disk model are models of the hyperbolic plane. An advantage of the Poincaré disk model is that it is conformal (circles and angles are not distorted); a disadvantage is that lines of the geometry are circular arcs orthogonal to the boundary circle of the disk.

The two models are related through a projection on or from the hemisphere model. The Klein model is an orthographic projection to the hemisphere model while the Poincaré disk model is a stereographic projection.

When projecting the same lines in both models on one disk both lines go through the same two ideal points. (the ideal points remain on the same spot) also the pole of the chord is the centre of the circle that contains the arc.

If P is a point a distance s from the centre of the unit circle in the Beltrami–Klein model, then the corresponding point on the Poincaré disk model a distance of u on the same radius:

$$u = \frac{s}{1+\sqrt{1-s^2}} = \frac{\left(1-\sqrt{1-s^2}\right)}{s}.$$

Conversely, If P is a point a distance u from the centre of the unit circle in the Poincaré disk model, then the corresponding point of the Beltrami–Klein model is a distance of s on the same radius:

$$s = \frac{2u}{1+u^2}.$$

Relation of the Disk Model to the Hyperboloid Model

Both the hyperboloid model and the Klein disk model are models of the hyperbolic plane.

The Klein disk (K, in the picture) is a gnomonic projection of the hyperboloid model (Hy) with as center the center of the hyperboloid (O) and the projection plane tangent to the nearest point of the hyperboloid.

Distance and Metric Tensor

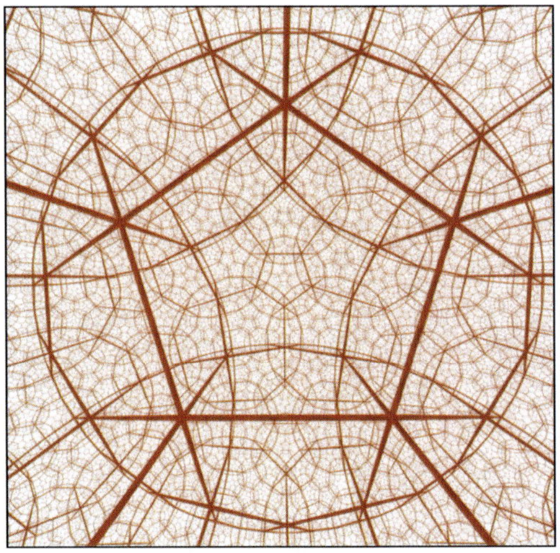

The regular hyperbolic dodecahedral honeycomb.

Given two distinct points U and V in the open unit ball of the model in Euclidean space, the unique straight line connecting them intersects the unit sphere at two ideal points A and B, labeled so that the points are, in order along the line, A, U, V, B. Taking the centre of the unit ball of the model as the origin, and assigning position vectors **u**, **v**, **a**, **b** respectively to the points U, V, A, B, we have that that $\|\mathbf{a}-\mathbf{v}\| > \|\mathbf{a} - \mathbf{u}\|$ and $\|\mathbf{u}-\mathbf{b}\| > \|\mathbf{v}-\mathbf{b}\|$, where $\|\cdot\|$ denotes the Euclidean norm. Then the distance between U and V in the modelled hyperbolic space is expressed as:

$$d(\mathbf{u},\mathbf{v}) = \frac{1}{2}\log\frac{\|\mathbf{v}-\mathbf{a}\|\|\mathbf{b}-\mathbf{u}\|}{\|\mathbf{u}-\mathbf{a}\|\|\mathbf{b}-\mathbf{v}\|},$$

where the factor of one half is needed to make the curvature −1.

The associated metric tensor is given by,

$$ds^2 = g(\mathbf{x}, d\mathbf{x}) = \frac{\|d\mathbf{x}\|^2}{1-\|\mathbf{x}\|^2} + \frac{(\mathbf{x}\cdot d\mathbf{x})^2}{\left(1-\|\mathbf{x}\|^2\right)^2} = \frac{(1-\|\mathbf{x}\|^2)\|d\mathbf{x}\|^2 + (\mathbf{x}\cdot d\mathbf{x})^2}{\left(1-\|\mathbf{x}\|^2\right)^2}$$

Relation to the Hyperboloid Model

The hyperboloid model is a model of hyperbolic geometry within $(n + 1)$-dimensional Minkowski space. The Minkowski inner product is given by,

$$\mathbf{x}\cdot\mathbf{y} = x_0 y_0 - x_1 y_1 - \cdots - x_n y_n$$

and the norm by $\|\mathbf{x}\| = \sqrt{\mathbf{x}\cdot\mathbf{x}}$. The hyperbolic plane is embedded in this space as the vectors **x** with $\|\mathbf{x}\| = 1$ and x_0 (the "timelike component") positive. The intrinsic distance (in the embedding) between points **u** and **v** is then given by,

$$d(\mathbf{u},\mathbf{v}) = \operatorname{arcosh}(\mathbf{u}\cdot\mathbf{v})$$

This may also be written in the homogeneous form,

$$d(\mathbf{u},\mathbf{v}) = \operatorname{arcosh}\left(\frac{\mathbf{u}}{\|\mathbf{u}\|}\cdot\frac{\mathbf{v}}{\|\mathbf{v}\|}\right),$$

which allows the vectors to be rescaled for convenience.

The Beltrami–Klein model is obtained from the hyperboloid model by rescaling all vectors so that the timelike component is 1, that is, by projecting the hyperboloid embedding through the origin onto the plane $x_0 = 1$. The distance function, in its homogeneous form, is unchanged. Since the intrinsic lines (geodesics) of the hyperboloid model are the intersection of the embedding with planes through the Minkowski origin, the intrinsic lines of the Beltrami–Klein model are the chords of the sphere.

Relation to the Poincaré Ball Model

Both the Poincaré ball model and the Beltrami–Klein model are models of the n-dimensional hyperbolic space in the n-dimensional unit ball in $\mathbf{R}^n$. If u is a vector of norm less than one representing a point of the Poincaré disk model, then the corresponding point of the Beltrami–Klein model is given by,

$$s = \frac{2u}{1+u\cdot u}.$$

Conversely, from a vector s of norm less than one representing a point of the Beltrami–Klein model, the corresponding point of the Poincaré disk model is given by,

$$u = \frac{s}{1+\sqrt{1-s\cdot s}} = \frac{\left(1-\sqrt{1-s\cdot s}\right)s}{s\cdot s}.$$

Given two points on the boundary of the unit disk, which are traditionally called *ideal points*, the straight line connecting them in the Beltrami–Klein model is the chord between them, while in the corresponding Poincaré model the line is a circular arc on the two-dimensional subspace generated by the two boundary point vectors, meeting the boundary of the ball at right angles. The two models are related through a projection from the center of the disk; a ray from the center passing through a point of one model line passes through the corresponding point of the line in the other model.

Hyperboloid Model

In geometry, the hyperboloid model, also known as the Minkowski model or the Lorentz model (after Hermann Minkowski and Hendrik Lorentz), is a model of n-dimensional hyperbolic geometry in which points are represented by the points on the forward sheet S^+ of a two-sheeted hyperboloid in $(n+1)$-dimensional Minkowski space and m-planes are represented by the intersections of the $(m+1)$-planes in Minkowski space with S^+. The hyperbolic distance function admits a simple expression in this model. The hyperboloid model of the n-dimensional hyperbolic space is closely related to the Beltrami–Klein model and to the Poincaré disk model as they are projective models in the sense that the isometry group is a subgroup of the projective group.

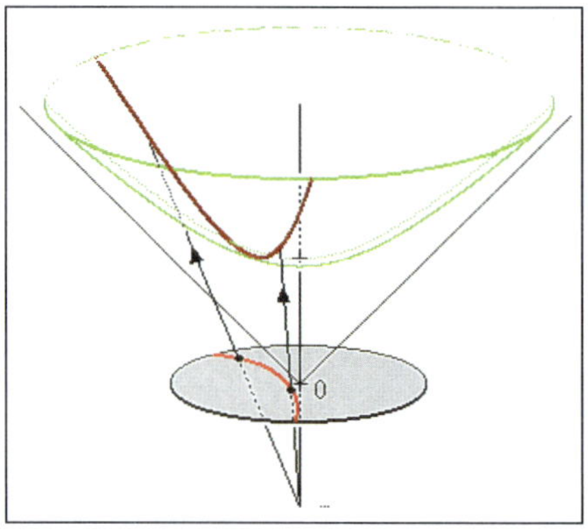

Red circular arc is geodesic in Poincaré disk model; it projects to the brown geodesic on the green hyperboloid.

Minkowski Quadratic Form

If $(x_0, x_1, \ldots, x_n)$ is a vector in the $(n+1)$-dimensional coordinate space $\mathbf{R}^{n+1}$, the Minkowski quadratic form is defined to be,

$$Q(x_0, x_1, \ldots, x_n) = x_0^2 - x_1^2 - \ldots - x_n^2.$$

The vectors $v \in R^{n+1}$ such that $Q(v) = 1$ form an n-dimensional hyperboloid S consisting of two connected components, or sheets: the forward, or future, sheet S^+, where $x_0 > 0$ and the backward, or past, sheet S^-, where $x_0 < 0$. The points of the n-dimensional hyperboloid model are the points on the forward sheet S^+.

The Minkowski bilinear form B is the polarization of the Minkowski quadratic form Q,

$$B(\mathbf{u}, \mathbf{v}) = (Q(\mathbf{u} + \mathbf{v}) - Q(\mathbf{u}) - Q(\mathbf{v})) / 2$$

Explicitly,

$$B((x_0, x_1, \ldots, x_n), (y_0, y_1, \ldots, y_n)) = x_0 y_0 - x_1 y_1 - \ldots - x_n y_n.$$

The hyperbolic distance between two points u and v of S^+ is given by the formula,

$$d(\mathbf{u}, \mathbf{v}) = \operatorname{arcosh}(B(\mathbf{u}, \mathbf{v})),$$

where arcosh is the inverse function of hyperbolic cosine.

Straight Lines

A straight line in hyperbolic n-space is modeled by a geodesic on the hyperboloid. A geodesic on the hyperboloid is the (non-empty) intersection of the hyperboloid with a two-dimensional linear subspace (including the origin) of the $n+1$-dimensional Minkowski space. If we take **u** and **v** to be basis vectors of that linear subspace with,

$$B(\mathbf{u},\mathbf{u}) = 1$$
$$B(\mathbf{v},\mathbf{v}) = -1$$
$$B(\mathbf{u},\mathbf{v}) = B(\mathbf{v},\mathbf{u}) = 0$$

and use w as a real parameter for points on the geodesic, then

$$\mathbf{u}\cosh w + \mathbf{v}\sinh w$$

will be a point on the geodesic.

More generally, a k-dimensional "flat" in the hyperbolic n-space will be modeled by the (non-empty) intersection of the hyperboloid with a $k+1$-dimensional linear subspace (including the origin) of the Minkowski space.

Isometries

The indefinite orthogonal group $O(1, n)$, also called the $(n + 1)$-dimensional Lorentz group, is the Lie group of real $(n + 1) \times (n + 1)$ matrices which preserve the Minkowski bilinear form. In a different language, it is the group of linear isometries of the Minkowski space. In particular, this group preserves the hyperboloid S. Recall that indefinite orthogonal groups have four connected components, corresponding to reversing or preserving the orientation on each subspace (here 1-dimensional and n-dimensional), and form a Klein four-group. The subgroup of $O(1, n)$ that preserves the sign of the first coordinate is the orthochronous Lorentz group, denoted $O^+(1, n)$, and has two components, corresponding to preserving or reversing the orientation of the spatial subspace. Its subgroup $SO^+(1, n)$ consisting of matrices with determinant one is a connected Lie group of dimension $n(n + 1)/2$ which acts on S^+ by linear automorphisms and preserves the hyperbolic distance. This action is transitive and the stabilizer of the vector $(1, 0,\ldots,0)$ consists of the matrices of the form,

$$\begin{pmatrix} 1 & 0 & \ldots & 0 \\ 0 & & & \\ \vdots & & A & \\ 0 & & & \end{pmatrix}$$

Where A belongs to the compact special orthogonal group $SO(n)$ (generalizing the rotation group $SO(3)$ for $n = 3$). It follows that the n-dimensional hyperbolic space can be exhibited as the homogeneous space and a Riemannian symmetric space of rank 1,

$$\mathbb{H}^n = SO^+(1,n)/SO(n)$$

The group $SO^+(1,n)$ is the full group of orientation-preserving isometries of the n-dimensional hyperbolic space.

References

- Non-euclidean-geometry, mathematics: encyclopedia.com, Retrieved 08 June, 2019
- Elliptic-space: encyclopedia2.thefreedictionary.com, Retrieved 25 May, 2019
- Quasi-elliptic-space: encyclopediaofmath.org, Retrieved 12 April, 2019
- Spherical-geometry: brilliant.org, Retrieved 19 July, 2019
- Hyperbolic-geometry, science: britannica.com, Retrieved 07 March, 2019
- Horocycles-congruent- 5510: faculty.etsu.edu, Retrieved 14 January, 2019
- Tessellations-by-Polygons: mathstat.slu.edu, Retrieved 26 February, 2019
- Gaussian-curvature: web.cs.iastate.edu, Retrieved 28 June, 2019

4

Understanding Discrete Geometry

Discrete geometry is the study of geometrical objects and properties that are discrete in their representation. Hyperplanes, polyhedrons, polytope, circle packing, Helly's theorem, etc. come within it. The topics elaborated in this chapter will help in gaining a better perspective about discrete geometry.

Discrete geometry is a mathematical discipline studying the properties of discrete geometric structures, which are arrangements of elementary geometric objects such as points, lines, planes, circles, spheres, polygons, polytopes, etc.; it also provides methods for solving various problems defined on such structures. Discrete geometry is closely related to computational geometry and other mathematical subjects, such as combinatorial optimization, geometric graph theory, combinatorial geometry and topology, finite geometry, and convex geometry. A specific branch of discrete geometry is the digital geometry which deals with "digital" objects, usually defined on integer lattices.

Discrete geometry has various applications in theoretical disciplines such as functional analysis, calculus of variations, geometry of numbers, geometric measure theory, group theory, algebraic geometry and topology, and mathematical physics. Results of discrete geometry have also been widely used in applied areas such as crystallography, tomography, and rigidity of surfaces. Various engineering applications of discrete geometry are found in coding theory, telecommunications, image processing, robot motion planning, and aerospace engineering. In particular, digital geometry has been developed with the explicit goal to provide rigorous mathematical foundations and basic algorithms for imaging sciences, including image analysis and processing, pattern recognition, computer vision and image understanding, biometrics, computer graphics, and medical imaging. These are in turn applicable to important and societally sensitive areas like medicine, defense, and security.

Arrangements of Hyperplanes

Let V be an I-dimensional affine space over the field K. An arrangement of hyperplanes, A, is a finite collection of codimension-one affine subspaces in V.

Examples of Arrangements of Hyperplanes

1. A subset of the coordinate hyperplanes is called a Boolean arrangement.
2. An arrangement is in general position if at each point it is locally Boolean.
3. The braid arrangement consists of the hyperplanes $\{x_i = x_j : 1 \leq i < j \leq 1\}$. It is the set of reflecting hyperplanes of the symmetric group on I letters.

4. The reflecting hyperplanes of a finite reflection group.

Combinatorics

An edge X of A is a non-empty intersection of elements of A. Let L (A) be the set of edges partially ordered by reverse inclusion. Then L is a geometric semi-lattice with minimal element V, rank given by codimension, and maximal elements of the same rank, $r(A)$. The Möbius function on L is defined by $\mu(V)$ and, for X > V,

$$\sum_{V \le Y \le X} \mu(Y) = 0.$$

The characteristic polynomial of A is $\mathcal{X}(A, t) = \sum_{X \in L} \mu(X)_t^{\text{codim X}}$. Let $\beta(A) = (-1)^{r(A)} X(A, 1)$. For a generic arrangement of n hyperplanes,

$$\mathcal{X}(A, t) = \sum_{k=0}^{r(A)} (-1)^k \binom{n}{k} t^{1-k}.$$

For the braid arrangement,

$$\mathcal{X}(A, t) = t(t-1)\ldots(t-(l-1)).$$

Similar factorizations hold for all reflection arrangements involving the (co)exponents of the reflection group. Given a P-tuple of hyperplanes, $S = (H_1, \ldots, H_p)$, let $\cap S = H_1 \cap \ldots \cap H_p$; note that $\cap S$ may be empty. S is dependent if $\cap S \ne \emptyset$ and $\text{codim}(\cap S) < |S|$. Let $E(A)$ be the exterior algebra on symbols (H) for $H \in A$, where the product is juxtaposition. Define $\partial : E \to E$ by $\partial 1 = 0$, $\partial(H) = 1$ and, for $p \ge 2$,

$$\partial(H_1 \ldots . H_p) = \sum_{k=1}^{p} (-1)^{k-1} (H_1 \ldots . \widehat{H_k} .. H_p).$$

Let $I(A)$ be the ideal of $E(A)$ generated by $\{S : \cap S = \phi\} \cup \{\partial S : S \text{ dependent}\}$. The Orlik–Solomon algebra of A is $A(A) = E(A) / I(A)$.

Divisor

The divisor of A is the union of the hyperplanes, denoted by $N(A)$. If $\mathbf{K} = \mathbf{R}$ or $\mathbf{K} = \mathbf{C}$, then $\boldsymbol{N}$ has the homotopy type of a wedge of $\beta(A)$ spheres of dimension $\boldsymbol{r}(A) - 1$. The singularities of N are not isolated. The divisor of a general-position arrangement has normal crossings, but this is not true for arbitrary A. Blowing-up N along all edges where it is not locally a product of arrangements yields a normal crossing divisor.

Complement

The complement of A is $M(A) = V \setminus N(A)$.

1. If $\mathbf{K} = \mathbf{F}_q$, then M is a finite set of cardinality $|M| = X(A, q)$.

2. If $\mathbf{K} = \mathbf{R}$, then M is a disjoint union of open convex sets (chambers) of cardinality $(-1)^1 X(A,-1)$. If $r(A)=1$, M contains $\beta(A)$ chambers with compact closure.
3. If **K** **C**, then M is an open complex (Stein) manifold of the homotopy type of a finite CW-complex. Its cohomology is torsion-free and its Poincaré polynomial (cf. Künneth formula) is $\text{Point}(M.t)=(-t)^1 X(A,-t^{-1})$. The product structure is determined by the isomorphism of graded algebras $H^*(M) \simeq A(A)$. The fundamental group of M has an effective presentation, but the higher homotopy groups of M are not known in general.

The complement of a Boolean arrangement is a complex torus. In a general-position arrangement of $n>1$ hyperplanes, M has non-trivial higher homotopy groups. For the braid arrangement, M is called the pure braid space and its higher homotopy groups are trivial. The symmetric group acts freely on M with as orbit space the braid space whose fundamental group is the braid group. The quotient of the divisor by the symmetric group is called the discriminant, which has connections with singularity theory.

Ball Quotients

Examples of algebraic surfaces whose universal cover is the complex ball were constructed as "Kummer" covers of the projective plane branched along certain arrangements of projective lines.

Logarithmic Forms

For $H \in A$, choose a linear polynomial αH with $H = \ker \alpha H$ and let $Q(A) = \Pi_{H \in A} \alpha H$. Let $\Omega^p[V]$ denote all global regular (i.e., polynomial) p-forms on V. Let $\Omega^p(V)$ denote the space of all global rational p-forms on V. The space $\Omega^p(A)$ of logarithmic p-forms with poles along A is,

$$\Omega^p(A) =$$
$$= \{\omega \in \Omega^p(V) : Q\omega \in \Omega^p[V], Q(d\omega) \in \Omega^{p+1}[V]\}.$$

The arrangement is free if ( ) is a free module over the polynomial ring. A free arrangement A has integer exponents $\{b_1,....b_1\}$, so that $X(A,t)=\Pi_{k=1}^{1}(t-b_k)$. Reflection arrangements are free. This explains the factorization of their characteristic polynomials.

Hypergeometric Integrals

Certain rank-one local system cohomology groups of M may be identified with spaces of hypergeometric integrals. If the local system is suitably generic, these cohomology groups may be computed using the algebra $A(A)$. Only the top cohomology group is non-zero, and it has dimension $\beta(A)$.

Arrangement of Lines

Let L be a set of n lines in the plane. The set L induces a subdivision of the plane that consists of vertices, edges, and faces. Some of the edges and faces are unbounded. This subdivision is usually referred to as the arrangement induced by L, and it is denoted by $A(L)$. An arrangement is called simple if no three lines pass through the same point and no two lines are parallel. The (combinatorial) complexity of an arrangement is the total number of vertices, edges, and faces of the arrangement. Arrangements of lines and their higherdimensional counterparts occur frequently in computational geometry. Often a problem that is defined on a set of points is dualized and turned into a problem on arrangements. This is done because the structure of a line arrangement is more apparent than the structure of a point set. A line through a pair of points in the primal plane, for instance, becomes a vertex in the dual arrangement—a much more explicit feature. The extra structure in an arrangement does not come for free: constructing a full arrangement is a time- and storage-consuming task, because the combinatorial complexity of an arrangement is high.

Theorem: Let L be a set of n lines in the plane, and let $A\ (L)$ be the arrangement induced by L.

- The number of vertices of $A\ (L)$ is at most n(n – 1)/2.
- The number of edges of $A\ (L)$ is at most n^2.
- The number of faces of $A\ (L)$ is at most $n^2/2+ n/2+1$.

Equality holds in these three statements if and only if $A\ (L)$ is simple.

Proof. The vertices of $A\ (L)$ are the intersection points of pairs of lines in L. Hence, there are at most n(n – 1)/2 of them. This number of vertices is achieved if and only if every pair of lines gives rise to a unique intersection point, which happens exactly if $A\ (L)$ is simple.

The number of edges on a line is one more than the number of vertices on that line. The latter number is at most n – 1, so the number of edges on a line is bounded by n. This gives at most n^2 edges in total, which is achieved if and only if $A\ (L)$ is simple.

To bound the number of faces of $A\ (L)$ we add the lines one by one and bound the increase in the number of faces at each step. Let $L:\{\ell_i,....\ell_n\}$. For 1 ≤ i ≤ n, define $L_i := \{\ell_1,....\ell_i\}$. What is the increase in the number of faces when we add i and go from $A\ (L_{i-1})$ to $A\ (L_i)$? Every edge on i splits a face of $A\ (L_{i-1})$ into two. Hence, the number of faces increase by the number of edges of $A\ (L_{i-1})$ on ℓ_i. The latter number is at most i, so the total number of faces is at most,

$$1+\sum_{i=1}^{n} i = n^2/2+n/2+1.$$

Again, this is achieved if and only if $A\ (L)$ is simple.

So the arrangement $A\ (L)$ induced by a set L of lines is a planar subdivision of at most quadratic complexity. The doubly-connected edge list seems a suitable way to store an arrangement; with this representation we can efficiently list the edges of a given face, step from one face to a neighboring one, and so on. A doubly-connected edge list, however, can only store bounded edges, and an arrangement also has a number of unbounded edges. Therefore we place a 180 large bounding box that encloses the interesting part of the arrangement, that is, a bounding box that contains all vertices of the arrangement in its interior. The subdivision defined by the bounding box plus the part of the arrangement inside it has bounded edges only and can be stored in a doubly-connected edge list.

A bounding box $B(L)$ that contains all vertices of $A\ (L)$ in its interior can easily be computed in quadratic time: compute all intersection points of pairs of lines, and choose the leftmost one, the rightmost one, the bottom one, and the top one. An axis-parallel rectangle that contains these four points contains all vertices of the arrangement.

The incremental algorithm adds the lines $\ell_1, \ell_2, ...\ell_n$ one after the other and updates the doubly-connected edge list after each addition. Let A_i denote the subdivision of the plane induced by the bounding box B(L) and the part of $A\ (\{\ell_1, ...\ell_i\})$ inside B(L). To add the line i, we must split the faces in A_{i-1} that are intersected by ℓ_i. We can find these faces by walking along ℓ_i from left to right, as follows. Suppose we enter a face f through an edge e. We walk along the boundary of f following $Next()$-pointers in the doubly-connected edge list until we find the half-edge of the edge e' where ℓ_i leaves f. We then step to the next face using the Twin()-pointer of that half-edge to reach the other half-edge for e in the doubly-connected edge list. This way we find the next face in time proportional to the complexity of f. It can also happen that we leave f through a vertex v. In this case we walk around v, visiting its incident edges, until we find the next face intersected by ℓ_i. The doubly-connected edge list allows us to do this in time proportional to the degree of v. Figure illustrates how we traverse the arrangement.

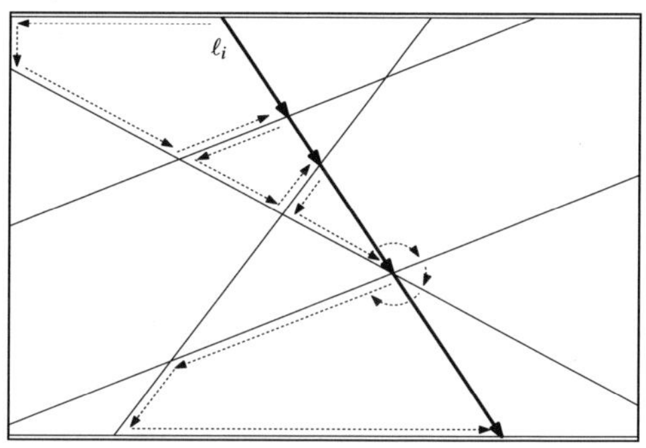

Traversing the arrangement.

Two things are left: how do we find the leftmost edge intersected by ℓ_i—this is the edge where we start the walk through A_{i-1}—and how do we actually split the faces we encounter?

The first issue is easy. The leftmost intersection point of i and A_{i-1} is an edge on B(L). We simply test all of them to locate the one where the traversal can be started. The face incident to this edge and inside B(L) is the first face that is split by ℓ_i. In case i intersects A_{i-1} first in a corner of B(L), the first face split by ℓ_i is the unique face incident to this corner and inside B(L). If ℓ_i is a vertical line we can locate the bottom intersection point of ℓ_i and A_{i-1} to start off the traversal. Since A_{i-1} contains at most $2i+2$ edges on B(L), the time needed for this step is linear for each line.

Suppose we have to split a face f, and assume that the face intersected by ℓ_i to the left of f has already been split. In particular, we assume that the edge e where we enter f has already been split. Splitting f is done as follows. First of all we create two new face records, one for the part of f above ℓ_i and one for the part of f below ℓ_i. Next we split e, the edge where i leaves f, and create a new vertex for $\ell_i \cap e'$. Thus we create one new vertex record, and two new half-edge records for both new edges. (If ℓ_i leaves f through a vertex, then this step is omitted.) Furthermore, we create half-edge records for the edge $\ell_i \cap$ f. It remains to correctly initialize the various pointers in the new face, vertex, and half-edge records, set some existing pointers to the new vertex record, half-edge records, and face records, and destroy the face record for f and the half-edge records for e'.

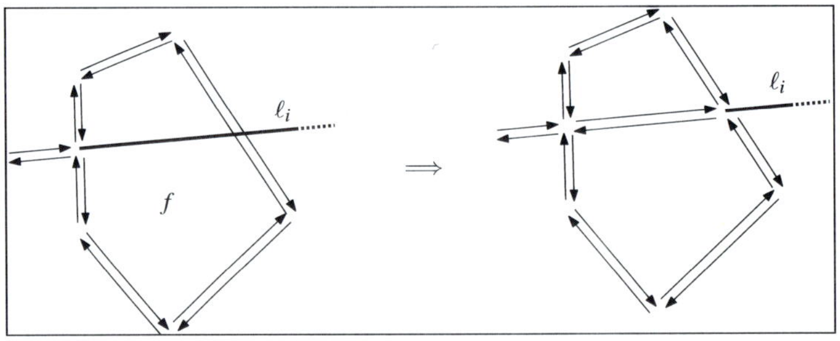

Splitting a face.

Polyhedrons

A solid shape bounded by polygons is called a polyhedron. The word polyhedral is the plural of word polyhedron.

- Faces: Polygons forming a polyhedron are known as its faces.
- Edges: Line segments common to intersecting faces of a polyhedron are known as its edges.
- Vertices: Points of intersection of edges of a polyhedron are known as its vertices.

In a polyhedron, three or more edges meet at a point to form a vertex. Following are some example of polyhedrons:

- Cuboids

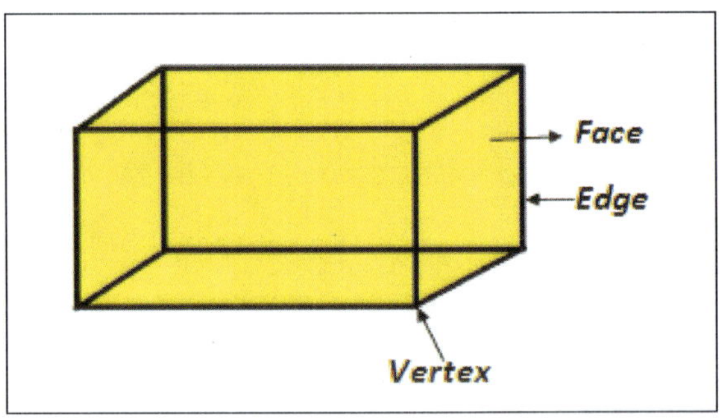

- Cube

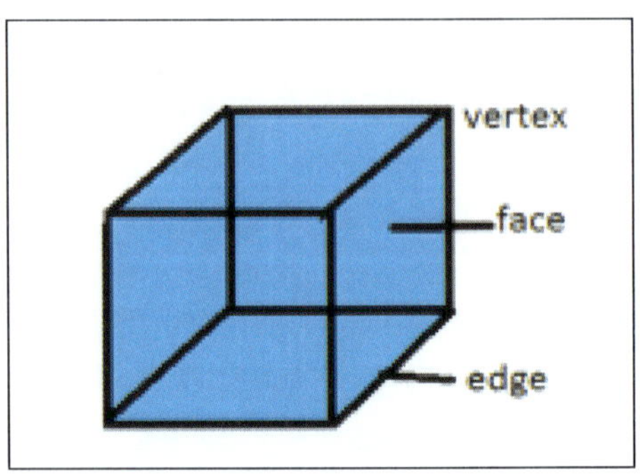

- Pyramid

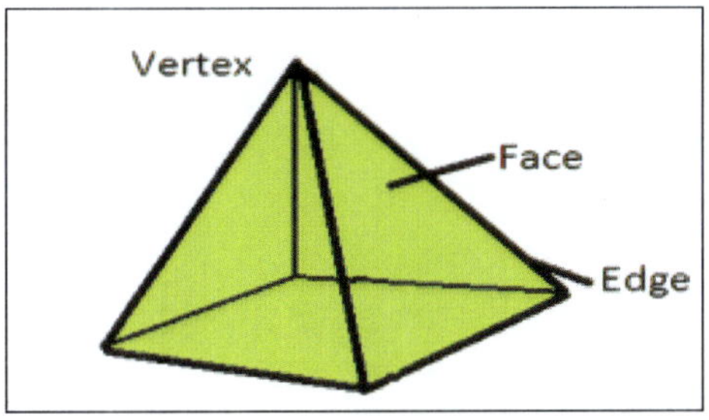

- Triangular pyramid

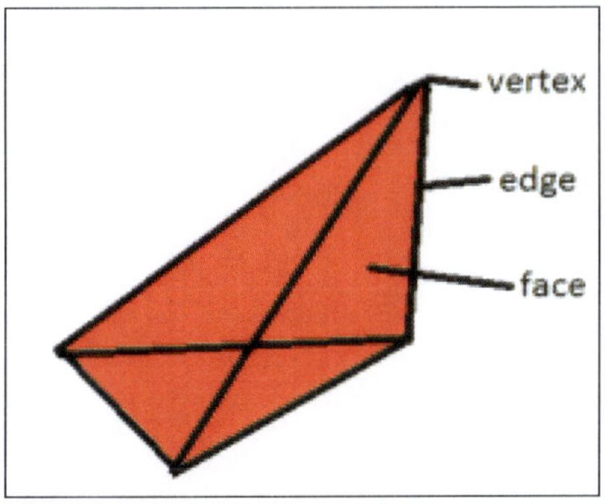

Regular Polyhedron

A polyhedron is said to be a regular polyhedron if its faces are made up of regular polygons and the same number of faces meet at each vertex. This means that the faces of a regular polyhedron are congruent regular polygons and its vertices are formed by the same number of faces. A cube is a regular polyhedron but a cuboid is not a regular polyhedron as its faces are not congruent rectangles.

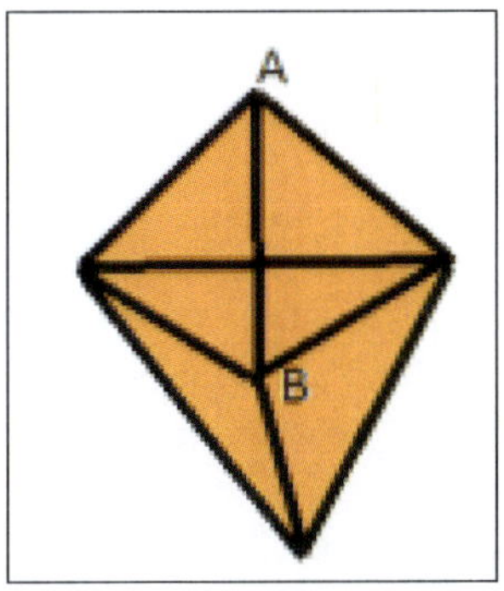

It is not regular because its faces are congruent triangles but the vertices are not formed by the same number of faces. Clearly, 3 faces meet at A but 4 faces meet at B.

Convex Polyhedron

If the line segment joining any two points on the surfaces of a polyhedron entirely lies inside or on the polyhedron, then it is said to be a convex polyhedron. Otherwise, it is known as the concave polyhedron.

Prisms and Pyramids

Two important members of polyhedron family are prism and pyramid. So, let us know more about these two polyhedrons.

Prism

A prism is a solid, whose side faces are parallelograms and whose bases are congruent parallel rectilinear figures.

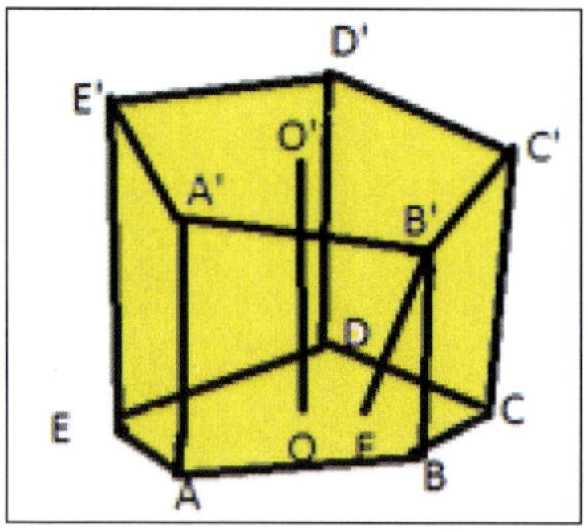

In figure, there is a prism whose bases are rectilinear figures ABCDE and A'B'C'D'E'.

- The Base of a Prism: The end on which a prism may be supposed to stand is called the base of the prism. In figure ABCDE and A'B'C'D'E' are the bases of the prism. Every prism has two bases.
- The Height of a Prism: The perpendicular distance between the ends of a prism is called the height of the prism. In figure B'F is the perpendicular distance between the ends ABCDE and A'B'C'D'E'. So it is the height of the prism shown in figure.
- Axis of prism: The straight line joining the centres of the ends of the prism is called the axis of the prism. In figure a straight line passing through O' and O is the axis of the prism.
- Length of prism: A length of a prism is a portion of the axis that lies between the parallel ends. In figure OO' is the length of the prism.
- Lateral faces: All faces other than the basis of a prism are known as its lateral faces. In figure, ABB'A', BCC'B', CDD'C', etc are lateral faces.
- Lateral edges: The lines of intersection of the lateral faces of a prism are called the lateral edges of the prism. In figure AA', BB', CC', DD' and EE' are the lateral edges of the prism.

Regular Prism

A prism is called a regular prism if its ends are a regular polygon.

Right Prism

A prism is called a right prism if its lateral edges are perpendicular to its ends (bases). Otherwise, it is said to be an oblique prism.

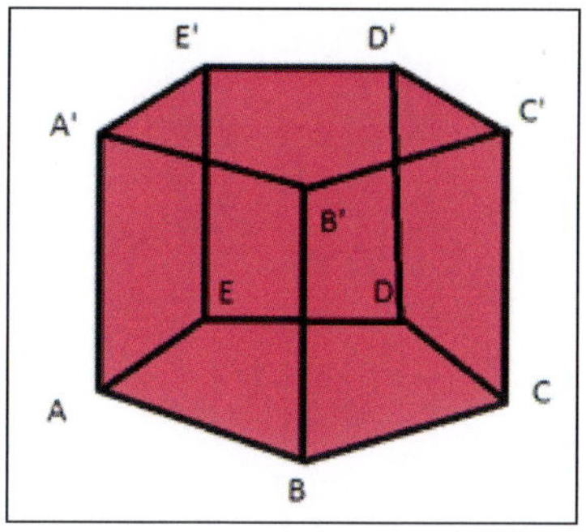

Triangular Prism

Prism is called a triangular prism if its ends are a triangle. A right prism is called a right triangular prism if its ends are triangles. In other words, a triangular prism is called a right triangular prism if its lateral edges are perpendicular to its ends.

A prism is said to be a quadrilateral prism or a pentagonal prism or a hexagonal prism, etc according to the number of sides in the rectilinear figure forming the ends is four or five or six, etc. If the ends of the quadrilateral prism are a parallelogram, then it is also known as a paralleled. A quadrilateral prism with its ends as squares is called a rectangular solid or cuboids.

Pyramid

A pyramid is a polyhedron whose base is a polygon of any number of sides and whose other faces are triangles with a common vertex.

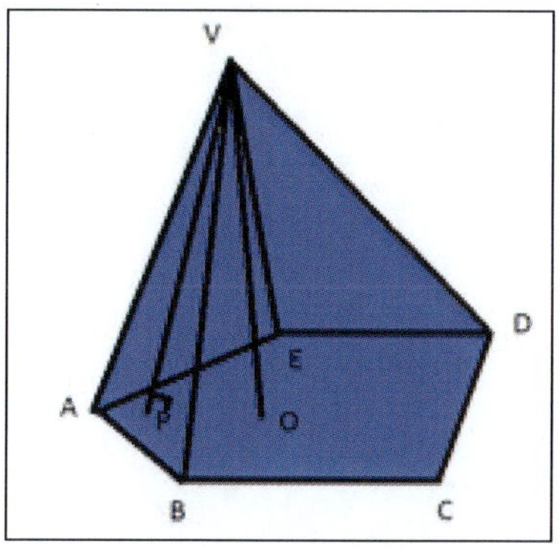

If all the corners of a polygon are joined to a point not lying in a plane, we get a pyramid. The figure shows a pyramid ABCDE. The base of this prism is the pentagon ABCDE and triangles VAB, VBC, VDE, and VEA are five faces.

- Vertex: The common vertex of the triangular faces of a pyramid is called the vertex oof the pyramid. In the figure V is the vertex of the pyramid.
- Height: The height of a pyramid is the length of the perpendicular from the vertex to the base. In the figure VP is the height of the pyramid.
- Axis: The axis of a pyramid is a straight line joining the vertex to the central point on the base. In figure VO is the axis of the pyramid.

- Lateral edges: The edges through the vertex of a pyramid are known as its lateral edges.
- Lateral faces: The side faces of a pyramid are known as its lateral faces.

Polytope

The word polytope is used to mean a number of related, but slightly different mathematical objects. A convex polytope may be defined as the convex hull of a finite set of points (which are always bounded), or as a *bounded* intersection of a finite set of half-spaces. Coxeter defines polytope as the general term of the sequence "point, line segment, polygon, polyhedron" or more specifically as a finite region of n-dimensional space enclosed by a finite number of hyperplanes. The special name polychoron is sometimes given to a four-dimensional polytope. However, in algebraic topology, the underlying space of a simplicial complex is sometimes called a polytope.

The part of the polytope that lies in one of the bounding hyperplanes is called a cell.

A *d*-dimensional polytope may be specified as the set of solutions to a system of linear inequalities,

$$m\mathbf{x} \leq \mathbf{b}$$

where **m** is a real $s \times d$ matrix and **b** is a real s-vector. The positions of the vertices given by the above equations may be found using a process called vertex enumeration.

A regular polytope is a generalization of the Platonic solids to an arbitrary dimension. The regular polytopes were discovered before 1852 by the Swiss mathematician Ludwig Schläfli. For n dimensions with $n \geq 5$, there are only three regular convex polytopes: the hypercube, cross polytope, and regular simplex, which are analogs of the cube, octahedron, and tetrahedron.

Convex Hulls

Intuitively, the convex hull captures the idea of points being inside other points. Imagine placing three thumbtacks on a board. Now wrap a piece of string around the thumbtacks until you return to where you started. The region now enclosed by the string is the convex hull of your three thumbtacks. Here's an example:

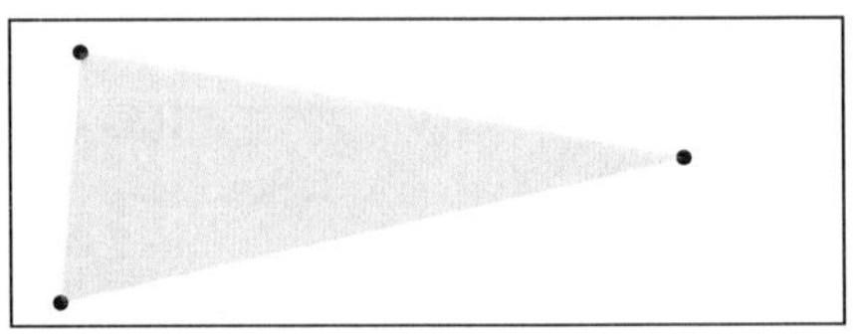

The convex hull (shaded) of the 3 black points.

The shaded triangle is the convex hull of the three black points. Now start with as many thumbtacks as you like, and wrap a string around the outside. Again, the region enclosed by the string is the convex hull of your thumbtacks. Another way to construct a polytope is by wrapping a rubber band around your points and letting it shrink. The enclosed region is again your polytope.

Here's an example with more points:

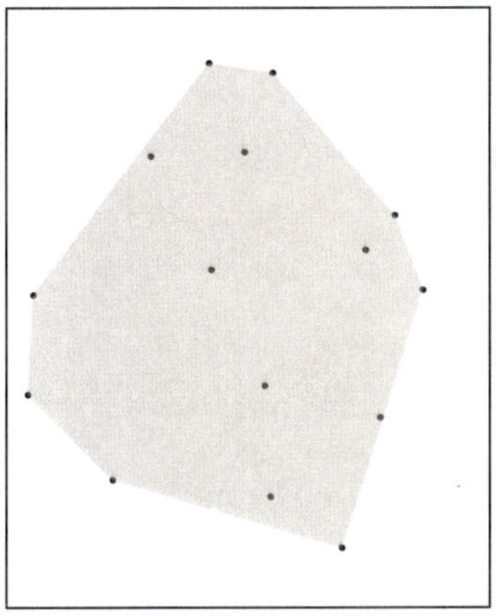

The convex hull (shaded) of the 15 black points.

Here, we have a polytope with more sides than before, and some points landed inside the shaded region. This means that they were not necessary to define the polytope.

Now we can keep going. Instead of just adding more points however, we'll add more dimensions too. In higher dimensions, imagine picking some points, and then putting a stretchy rubber suit around those points. Eventually, as the rubber shrinks aruond your points, you'll see the convex hull as the region enclosed by the rubber suit. For example, the 3-cube is the convex hull of the points: (0, 0, 0),(0, 0, 1),(0, 1, 0),(1, 0, 0),(0, 1, 1),(1, 0, 1),(0, 1, 1),(1, 1, 1).

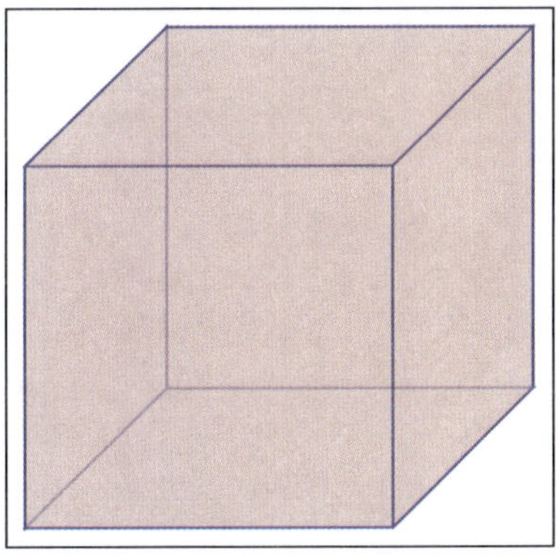

The 3-cube is the convex hull of its corners.

Faces

Now that we have a rough idea about polytopes, we can study some of their properties. A textbook definition of a face of a polytope goes as follows: A face of a d-polytope P is a subset $F \subset P$ maximized by some linear functional on $\mathbb{R}^d$. However, as before, we can define a face much more geometrically.

Imagine dropping a cube onto a flat surface. The points of the cube that touch the ground form a face of the cube. Most likely, the cube will fall on one of its 8 square sides, so these squares are faces of the cube. Less likely, the cube could have fallen on one of its 12 edges, so these edges are also faces of the cube. Least likely, the cube could have fallen onto one of its corners, so these corners are also faces of the cube.

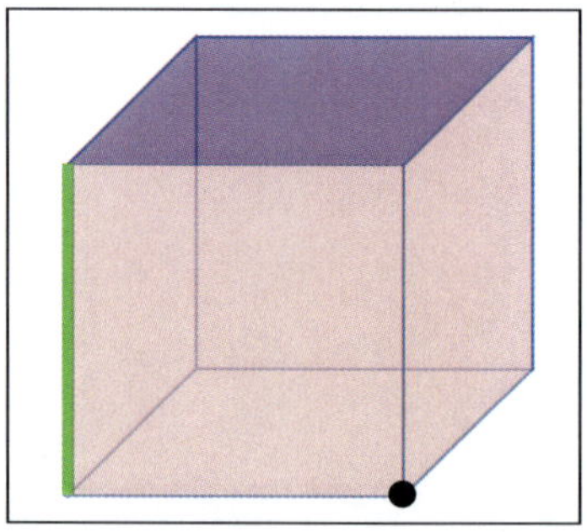

The blue square, green edge, and black vertex are all faces of the 3-cube.

More generally, for any polytope, faces can be thought of as the sets of points that can rest on a flat surface. For example, no pair of opposite sides of a cube form a face, because there is no way the top and bottom of a cube can be placed simultaneously on the same surface.

Circle Packing

A circle packing is an arrangement of circles inside a given boundary such that no two overlap and some (or all) of them are mutually tangent. The generalization to spheres is called a sphere packing. Tessellations of regular polygons correspond to particular circle packings. There is a well-developed theory of circle packing in the context of discrete conformal mapping (Stephenson).

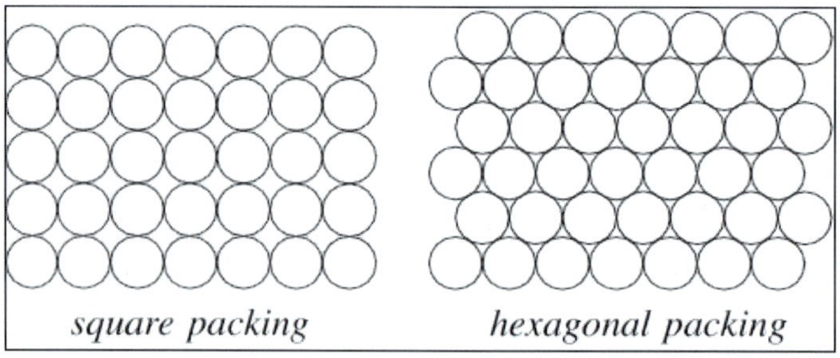

The densest packing of circles in the plane is the hexagonal lattice of the bee's honeycomb, which has a packing density of:

$$\eta_h = \frac{1}{6}\pi\sqrt{3} \approx 0.9068996821$$

Gauss proved that the hexagonal lattice is the densest plane lattice packing, and in 1940, L. Fejes Tóth proved that the hexagonal lattice is indeed the densest of all possible plane packings.

Surprisingly, the circular disk is not the least economical region for packing the plane. The "worst" packing shape is not known, but among centrally symmetric plane regions, the conjectured candidate is the so-called smoothed octagon.

Wells considers the maximum size possible for n identical circles packed on the surface of a unit sphere.

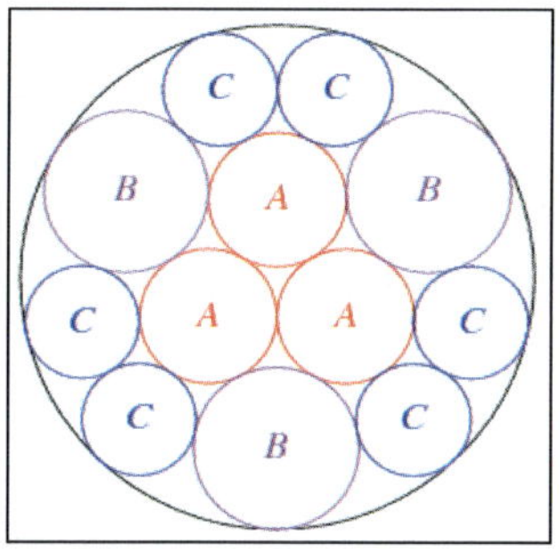

Using discrete conformal mapping, the radii of the circles in the above packing inside a unit circle can be determined as roots of the polynomial equations,

$$a^6 + 378a^5 + 3411a^4 - 8964\ a^3 - 10233\ a^2 + 3402\ a - 27 = 0$$
$$169\ b^6 + 24978\ b^5 + 2307\ b^4 - 14580\ b^3 + 3375\ b^3 + 3375\ b^2\ 162b - 27 = 0$$
$$c^6 + 438\ c^5 + 19077\ c^4 - 15840\ c^3 - 360c^2 + 2592c - 432 = 0$$

With,

$$a \approx 0.266746$$
$$b \approx 0.321596$$
$$c \approx 0.223138.$$

The following table gives the packing densities η for the circle packings corresponding to the regular and semiregular plane tessellations.

tessellation	η exact	η approx.
$\{3,6\}$	$\frac{1}{12}\sqrt{12}\ \pi$	0.9069
$\{4,4\}$	$\frac{1}{4}\pi$	0.7854
$\{6,3\}$	$\frac{1}{9}\sqrt{3}\ \pi$	0.6046
$3^2\ .4^2$	$\left(2-\sqrt{3}\right)\pi$	0.8418
$3^2\ .4\ .3\ .4$	$\left(2-\sqrt{3}\right)\pi$	0.8418
3.6 .3 .6	$\frac{1}{8}\sqrt{3}\ \pi$	0.6802
$3^4\ .6$	$\frac{1}{7}\sqrt{2}\ \pi$	0.7773

3.12^2	$\left(7\sqrt{3}-12\right)\pi$	0.3907
4.8^2	$\left(3-2\sqrt{2}\right)\pi$	0.5390
3.4 .6 .4	$\frac{1}{2}\left(2\sqrt{3}-3\right)\pi$	0.7290
4.6 .12	$\frac{1}{3}\left(2\sqrt{3}-3\right)\pi$	0.4860

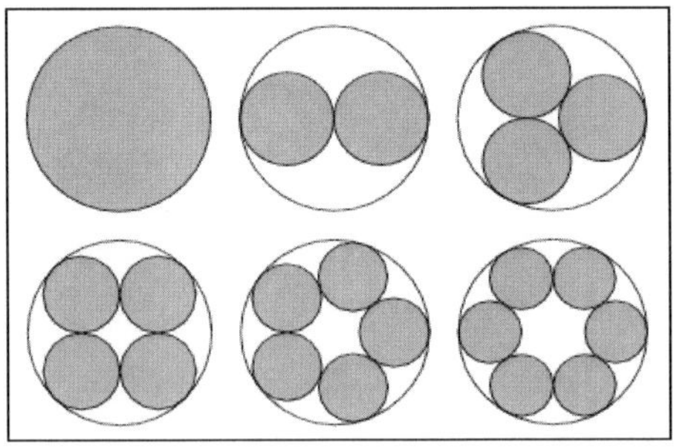

Solutions for the smallest diameter circles into which n unit-diameter circles can be packed have been proved optimal for n = 1 through 10. The best known results are summarized in the following table, and the first few cases are illustrated above (Friedman).

n	d_{exact}	d_{approx}
1	1	1.00000
2	2	2.00000
3	$1+\frac{2}{3}\sqrt{3}$	2.15470...
4	$1+\sqrt{2}$	2.41421...
5	$1+\sqrt{2\left(1+1/\sqrt{5}\right)}$	2.70130...
6	3	3.00000
7	3	3.00000
8	$1+\csc\left(\pi/7\right)$	3.30476...
9	$1+\sqrt{2\left(2+\sqrt{2}\right)}$	3.61312...
10		3.82...
11		
12		4.02...

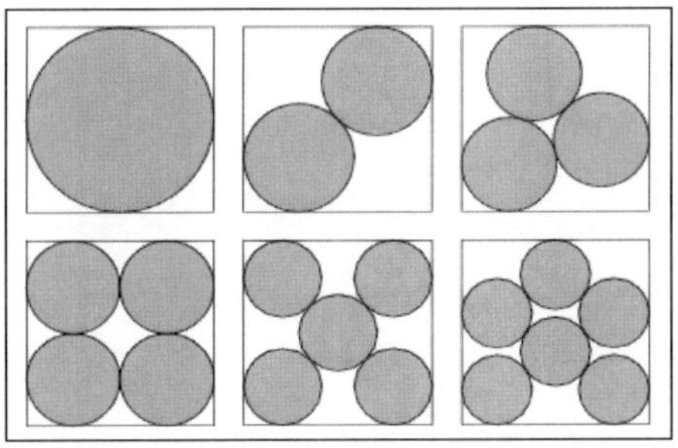

The following table gives the diameters d of circles giving the densest known packings of n equal circles packed inside a unit square, the first few of which are illustrated above (Friedman). All $n=1$ to 20 solutions (in addition to all solutions $n=k^2$) have been proved optimal (Friedman). Peikert uses a normalization in which the centers of n circles of diameter m are packed into a square of side length 1. Friedman lets the circles have unit radius and gives the smallest square side length s. A tabulation of analytic s and diagrams for $n=1$ to 25 circles is given by Friedman. Coordinates for optimal packings are given by Nurmela and Östergård.

n	d	$\approx d$	m	$\approx m$
1	1	1.000000		
2	$\frac{2}{2+\sqrt{2}}$	0.585786	$\sqrt{2}$	1.414214
3	$\frac{4}{4+\sqrt{2}+\sqrt{6}}$	0.508666	$\sqrt{6}-\sqrt{2}$	1.035276
4	$\frac{1}{2}$	0.500000	1	1.000000
5	$\sqrt{2}-1$	0.414214	$\frac{1}{2}\sqrt{2}$	0.707107
6	$\frac{1}{23}\left(6\sqrt{13}-13\right)$	0.375361	$\frac{1}{6}\sqrt{3}$	0.600925
7	$\frac{2}{13}\left(4-\sqrt{3}\right)$	0.348915	$4-2\sqrt{3}$	0.535898
8	$\frac{2}{2\sqrt{2}+\sqrt{6}}$	0.341081	$\frac{1}{2}\left(\sqrt{6}-\sqrt{2}\right)$	0.517638
9	$\frac{1}{3}$	0.333333	$\frac{1}{2}$	0.500000
10		0.296408		0.421280

The smallest square into which two unit circles, one of which is split into two pieces by a chord, can be packed is not known.

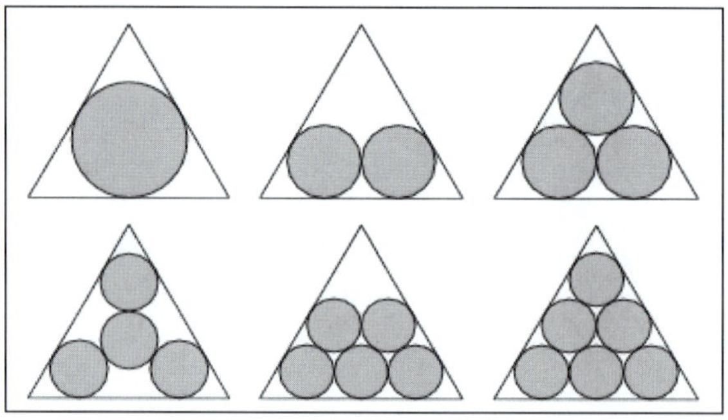

The best known packings of circles into an equilateral triangle are shown above for the first few cases.

Cubic Close Packing

There are three types of cubic lattices corresponding to three types of cubic close packing, as summarized in the following table. Now that the Kepler conjecture has been established, hexagonal close packing and face-centered cubic close packing, both of which have packing density of $\eta = \pi / \left(3\sqrt{2}\right) = 0.74048...$, are known to be the densest possible packings of equal spheres.

Lattice type	Basis vectors	Packing density
Simple cubic (SC)	$\hat{\mathbf{x}}, \hat{\mathbf{y}}, \hat{\mathbf{z}}$	$\pi / 6 \approx 52.3\%$
Face-centered cubic (FCC)	$\frac{1}{2}\left(\hat{\mathbf{y}}+\hat{\mathbf{z}}\right), \frac{1}{2}\left(\hat{\mathbf{x}}+\hat{\mathbf{z}}\right), \frac{1}{2}\left(\hat{\mathbf{x}}+\hat{\mathbf{y}}\right)$	$\pi / \left(3\sqrt{2}\right) \approx 74.0\%$
Body-centered cubic (BCC)	$\frac{1}{2}\left(-\hat{\mathbf{x}}+\hat{\mathbf{y}}+\hat{\mathbf{z}}\right), \frac{1}{2}\left(\hat{\mathbf{x}}-\hat{\mathbf{y}}+\hat{\mathbf{z}}\right), \frac{1}{2}\left(\hat{\mathbf{x}}+\hat{\mathbf{y}}-\hat{\mathbf{z}}\right)$	$\pi\sqrt{3} / 8 \approx 68.0\%$

Simple cubic packing consists of placing spheres centered on integer coordinates in Cartesian space.

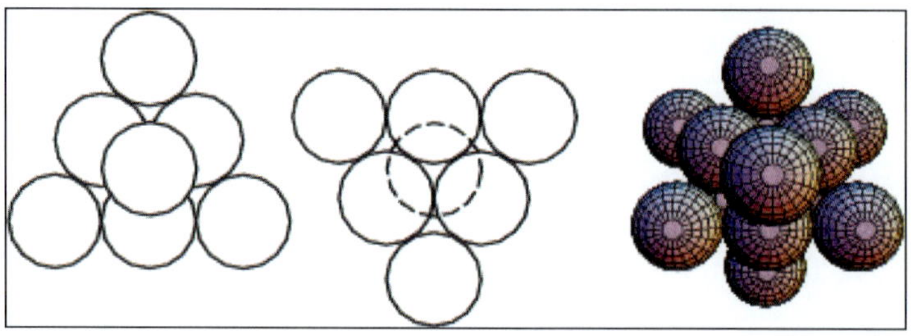

Arranging layers of close-packed spheres such that the spheres of every third layer overlay one another gives face-centered cubic packing. To see where the name comes from, consider packing six spheres together in the shape of an equilateral triangle and place another sphere on top to create a triangular pyramid. Now create another such grouping of seven spheres and place the two pyramids together facing in opposite directions.

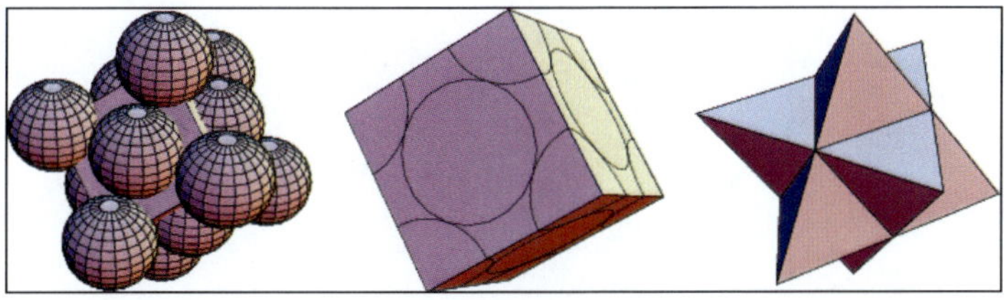

Connecting the centers of eight of the spheres, a cube emerges in which the centers of the other six spheres lie at the centers of the faces of the cube. Connecting the centers of these 14 spheres gives a stella octangula, illustrated above.

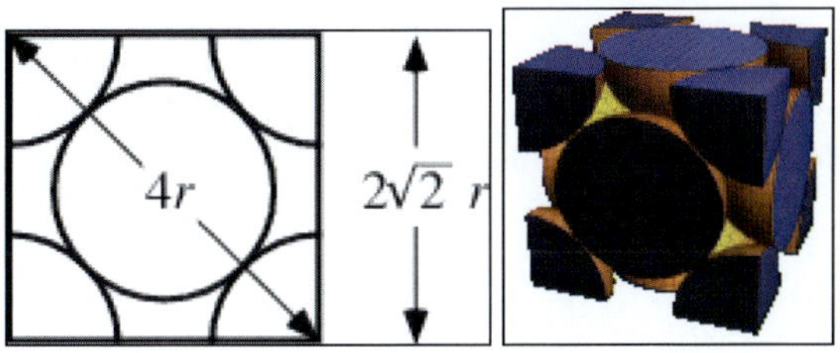

Consider the cube defined by 14 spheres in face-centered cubic packing. This "unit cell," one face of which is illustrated above in schematic form, contains eight 1/8-spheres (one at each polygon vertex) and six hemispheres. The total volume of spheres in the unit cell is therefore,

$$V_{\text{FCC spheres}} = \left(8\cdot\frac{1}{8}+6\cdot\frac{1}{2}\right)\frac{4\pi}{3}r^3 = \frac{16}{3}\pi r^3.$$

The diagonal of a face of the unit cell is $4r$, so each side is of length $2\sqrt{2}\,r$. The volume of the unit cell is therefore,

$$V_{\text{FCC unit cell}} = \left(2\sqrt{2}\,r\right)^3 = 16\sqrt{2}\,r^3,$$

giving a packing density $\eta = V_{\text{spheres}} / V_{\text{cell}}$ of,

$$\eta_{FCC} = \frac{\pi}{3\sqrt{2}} = 0.74048\ldots$$

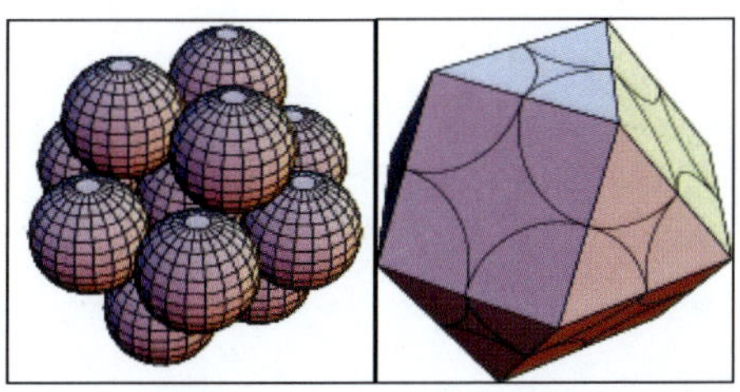

In face-centered cubic packing, each sphere is surrounded by 12 other spheres. Taking a collection of 13 such spheres gives the cluster illustrated above. Connecting the centers of the external 12 spheres gives a cuboctahedron.

In cubic body-centered packing, each sphere is surrounded by eight other spheres. The figure above shoes the unit cell in body-centered cubic packing. In this configuration, a single full sphere occupies the center and is surrounded by eight 1/8-spheres. The total volume of spheres in the unit cell is therefore,

$$V_{\text{BCC spheres}} = \left(8 \cdot \frac{1}{8} + 1\right)\frac{4\pi}{3}r^3$$
$$= \frac{8}{3}\pi r^3.$$

The space diagonal of the unit cell is $4r$, so each side is of length $4r/\sqrt{3}$. The volume of the unit cell is therefore,

$$V_{\text{BCC unit cell}} = \left(\frac{4r}{\sqrt{3}}\right)^3 = \frac{64}{3\sqrt{3}}r^3,$$

giving a packing density $\eta = V_{spheres} / V_{cell}$ of,

$$\eta_{\text{BCC}} = \frac{\sqrt{3}\,\pi}{8}$$
$$= 0.680174...$$

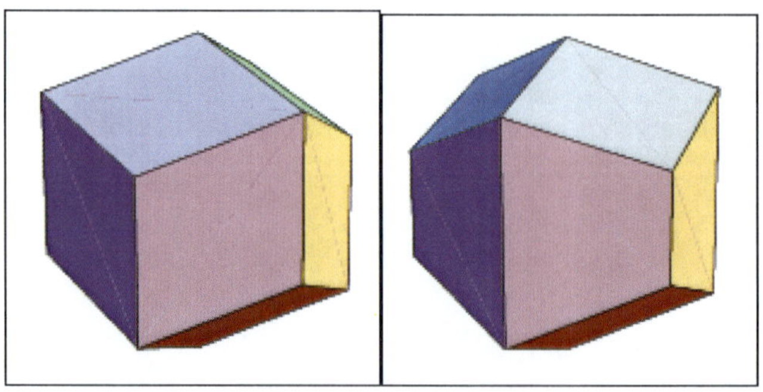

If spheres packed in a cubic lattice, face-centered cubic lattice, and hexagonal lattice are allowed to expand uniformly until running into each other, they form cubes, hexagonal prisms, and rhombic dodecahedra, respectively. In particular, if the spheres of face-centered cubic packing are expanded until they fill up the gaps, they form a solid rhombic dodecahedron (left figure above), and if the spheres of hexagonal close packing are expanded, they form a second irregular dodecahedron consisting of six rhombi and six trapezoids known as the trapezo-rhombic dodecahedron. The latter can be obtained from the former by slicing in half and rotating the two halves 60° with respect to each other. The lengths of the short and long edges of the rotated dodecahedron have lengths 2/3 and 4/3 times the length of the rhombic faces. Both the rhombic dodecahedron and trapezo-rhombic dodecahedron are space-filling polyhedra.

Helly's Theorem

Helly's theorem is a result from combinatorial geometry that explains how convex sets may intersect each other. The theorem is often given in greater generality, though for our considerations, we will mainly apply it to the plane.

We begin with a definition of a convex set. A set X is convex, if for every pair of points p and q in XX, the line segment $\overline{pq}$ lies within X. Note that the empty set is trivially a convex set, since there are no points within it.

Lemma: The intersection of two convex sets is convex.

Proof: Consider two convex sets X_1 and X_2. Let $X = X_1 \cap X_2$. Take any 2 points p and q in X. These points lie in X_1 (and in X_2). By definition of convex, the line segment $\overline{pq}$ lies in X_1 (and in X_2). Hence, $\overline{pq}$ lies in their intersection, which is X. Thus, X_1 is convex.

For those who are familiar with convex functions, we say that a function $f(x)$ is convex if the graph and the region above it, is a convex set. This is one way to remember the difference between a convex and concave function.

Statement of the Theorem

Helly's Theorem: Suppose $X_1, X_2, \ldots\ldots X_n$ is a collection of convex sets in $\mathbb{R}^d$, with $n > d$. If the intersection of every $d+1$ of these sets is non-empty, then the entire collection has a non-empty intersection.

This is a very surprising result. Knowledge of a local property (every d + 1 of the sets intersect) implies a global property (all of these sets must intersect), which is extremely rare.

We will prove the theorem in the case d = 2 (i.e., the plane). The general case proceeds in a similar manner and the reader is invited to flesh out the details. In the Worked Examples, we give a different way of approaching d = 1, though that method doesn't easily extend to higher dimensions.

We first prove a Lemma, which is Helly's Theorem for $d = 2,\ n = d + 2$.

Lemma: Given 4 convex subsets of the plane, if every 3 of them have a non-empty intersection, then all 4 of them have a non-empty intersection.

Proof: Let the subsets be X_1, X_2, X_3, X_4. Define a_i to be a common point of the three sets which do not include X_i. Now, consider the configuration of these 4 points.

If the convex hull of these 4 points is a triangle, WLOG let a_1 be in the interior. Then since a_2, a_3, a_4 is within X_1, hence a_1 is within the triangle given by a_2, a_3, a_4 and hence is within X_1. So a_1 is in all four sets.

If the convex hull of these 4 points is a convex quadrilateral, WLOG let the points be ordered as a_1, a_2, a_3, a_4. Now consider the diagonals $\overline{a_1a_3}$ and $\overline{a_2\,a_4}$. Since it is a convex quadrilateral, they intersect at a point p. Then, since $p \in \overline{a_1a_3}$, hence p is in sets X_2 and X_4. Similarly, since $p \in \overline{a_2a_4}$, hence p is in sets is in set X_1 and X_3. Thus, p is in all four sets.

We are now ready to prove Helly's Theorem in the plane.

Proof: We proceed by induction, in a slightly tricky manner. The base case $n = 3$ is trivially true. The base case $n = 4$ is the above Lemma.

For the induction step, suppose the theorem holds for some k. Given any set of k + 1 subsets $X_1, X_2, \ldots X_k, X_{k+1}$, consider the set of k subsets $Y_1 = X_1 \cap X_{k+1}, Y_2 = X_2 \cap X_{k+1}, \ldots, Y_k = X_k \cap X_{k+1}$. From the intersection property, we know that Y_i is a convex set for all i.

Now, consider the intersection of any 3 sets. We have,

$$Y_p \cap Y_q \cap Y_r = X_p \cap X_q \cap X_r \cap X_{k+1},$$

Which is nonempty by the Lemma. Applying the induction hypothesis to the sets Y_i, we conclude that the intersection of all sets Y_i is non-empty. But since $\cap X_i = \cap Y_i$, it follows that the interesc-tion of X_i is also non-empty.

The requirement of a finite collection of subsets is necessary. For example, on the number line, the collection of sets $\{[n, \infty)\}$ satisfies the property that any 2 of them have a non-empty intersection, but the intersection of all of these sets is empty. However, for an infinite collection of compact, convex subsets, Helly's Theorem still holds.

Worked Examples

Prove Helly's Theorem for d = 1, i.e., the case of the line.

Solution: The only convex sets on the line are line segments. To each of these sets, we will colour the left most point red, and the right most point blue. Then, the condition of intersection tells us that the red points must all occur to the left of the blue point.

We now consider the right most red point, and the left most blue point. If these 2 points are the same, then this point must lie in all of our sets. If these 2 points are distinct, then the line segment connecting these 2 points must lie in all of our sets. Hence we are done.

Note: As you can see, this is a straightforward argument, especially when compared with the above proof of Helly's Theorem. However, it doesn't extend to higher dimensions.

Find a collection of 3 sets in the plane ($d = 2$), such that every 2 of them intersect, but all 3 of them do not intersect.

Solution: Let ABC be a non-degenerate triangle. Let the sets be the line segments AB, BC, CA. Clearly, any 2 of them intersect (at a vertex), and the 3 of them do not intersect.

Note: This shows that the requirement that every d + 1 sets have a non-empty intersection is strict. Do you see how to generalize this counter example to all d?

Let $x_1, x_2, \ldots x_n$ be a finite set of points in the plane, such that the distance between any 2 points is smaller than 1. Show that there exists a disk of radius 1 which covers all of these points.

Solution: Let X_i be disks of radius 1 with center x_i. Then, every 3 of these disks X_i, X_j, and X_k have a non-empty intersection, since we know that the point x_i lies in $X_i \cap X_j \cap X_k$. Hence, by Helly's Theorem, there is a point x which lies in $\cap X_i$, which means that the distance between x and x_i is less than 1. Hence, the disk X of radius 1 with center x covers all of these points.

References

- Discrete-geometry-and-topology-and-their-applications-to-imaging-sciences, special-issue: guide2research.com, Retrieved 08 August, 2019
- Arrangement-of-hyperplanes: encyclopediaofmath.org, Retrieved 04 July, 2019
- Polyhedrons, maths-visualising-solid-shapes: toppr.com, Retrieved 10 June, 2019
- Polytope: mathworld.wolfram.com, Retrieved 15 May, 2019
- CirclePacking: mathworld.wolfram.com, Retrieved 16 April, 2019
- Hellys-theorem: brilliant.org, Retrieved 14 June, 2019
- CubicClosePacking: mathworld.wolfram.com, Retrieved 18 March, 2019

5

Algebraic Geometry

Algebraic geometry is a branch of mathematics which uses algebraic techniques for solving geometrical problems for set of zeros. Some of the concepts related to it are affine variety, Pascal's theorem, Reimann-Roch theorem, etc. This chapter closely examines about these concepts of algebraic geometry to provide an extensive understanding of the subject.

Algebraic geometry deals with curves or surfaces (or more abstract generalisations of these) which can be viewed both as geometric objects and as solutions of algebraic (specifically, polynomial) equations. A simple example is a circle of radius 1, which is the set of all points which are at a unit distance from its centre, but it is also the set of points (x, y) satisfying $x^2+y^2-1=0$. What geometric properties can be inferred from the equation? What about more complicated polynomial equations such as the semicubical parabola $y^2-x^3=0$, which has a cusp at its tip (such "singularities" are of great importance), or something less intelligible such as $x^2-x^2y^2+y^3+xy-1=0$? And, intriguingly, what if we change the minus sign in the equation for a circle to a plus, so that it reads $x^2+y^2+1=0$? Now x and y have to be complex, or there are no solutions at all.

Algebraic geometry sets out to answer these questions by applying the techniques of abstract algebra to the set of polynomials that define the curves (which are then called "algebraic varieties"). The mathematics involved is inevitably quite hard, although it is covered in degree-level courses. Other common questions in algebraic geometry concern points of special interest such as singularities, inflection points and points at infinity - we shall see these throughout the catalogue. More advanced questions in algebraic geometry concern relations between curves given by different equations and the topology of curves, and many other topics.

Singularities

Consider a hyperboloid of one sheet, a cone, and a hyperboloid of two sheets (shown below). Imagine 'morphing' between the one-sheet and two-sheet hyperboloids - the point at which we pass from one case to the other is the cone. On the cone, there is a singularity in the centre. Studying singularities is a crucial tool in geometry, as they occur at the 'boundary' of two different forms of behaviour. For example, the number of components of the surface changes here from one to two.

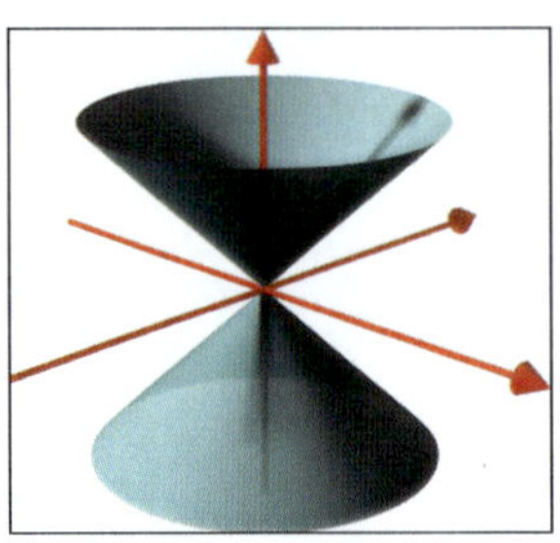

Surfaces defined by algebraic equations are "smooth" at almost all their points - by this we mean that, if you look closely enough, the surface looks like a plane $\mathbb{R}^2$. A "singularity" (or "singular point") of an algebraic surface f(x, y, z) = 0 is a point at which it is not smooth, for example having a 'cusp' or a 'ridge'.

In a small neighbourhood of a singular point, the surface fails to resemble $\mathbb{R}^2$. This can be made more precise using topology: the surface does not locally resemble a manifold at singular points.

It can also be helpful to look at just the real points on a surface, i.e. just those with coordinates in R, satisfying the equation of the surface. We will speak of singularities as being "real" if they are singularities of the surface defined over the reals, and "complex" otherwise.

Rings, Ideals and Polynomial Rings

Rings are a very powerful and useful object of study, especially within the topic of commutative algebra. A "ring" is a set together with two "operations", and can be thought of as a generalisation of the (positive and negative) integers with operations + and ×. Ring theory dates back to the 19th century, and has grown substantially with lots of mathematical topics, especially algebraic geometry, now relying on it.

An "ideal" is a special subset of a ring. An ideal is self-contained with respect to addition, and multiplying anything by a member of the ideal gives another member of the ideal. Intuitively, this generalises special subsets such as the even numbers within the integers. This term was coined by eminent German mathematician Richard Dedekind in the late 19th century, inspired by the notion of an ideal number.

A ring is "Noetherian" if all its ideals are finitely generated. This means that all members of an ideal can be found by "adding" and "multiplying", from some finite collection of starting elements. Working with such ideals is often desirable. Noetherian rings are named after Emmy Noether, who developed the foundation of modern commutative ring theory in 1921 with her landmark paper "Idealtheorie in Ringbereichen".

Given a ring R, its corresponding "polynomial ring" is the set of all 1 variable polynomials whose coefficients are members of R, and is denoted R[X]. For example, the integers $\mathbb{Z}$ have polynomial ring $\mathbb{Z}[X]$ containing elements such as x^2+3x-7 but not $x^3-\frac{1}{2}x+6$, which has a non-integer coefficient.

Algebraic Varieties and 2 Key Results

In algebraic geometry, the main objects of interest are "algebraic varieties", which are essentially geometric manifestations of solutions to polynomial equations. Some familiar examples include spheres, conic sections, and lines in $\mathbb{R}^2$ (two-dimensional Euclidean space). Algebraic surfaces can be thought of as particularly simple algebraic varieties, defined by only one polynomial equation. Our collection of mathematical models exhibits a range of interesting properties of algebraic surfaces.

Hilbert's Nullstellensatz, proven in 1890, is a remarkable theorem which helps establish the fundamental link between algebraic varieties and ideals in polynomial rings, thus unifying algebraic geometry and commutative algebra. Hilbert's Basis Theorem, also from 1890, says that for any

Noetherian ring, its "polynomial ring" is also Noetherian. The upshot of these results is a solid foundation for working with algebraic varieties.

Projective Space

A "field" is a special case of a ring, in which every non-zero element has a multiplicative inverse. This means that fields have a "division" operation analogous to what we have with real numbers. Indeed, the real numbers form a field. The integers do not, as $1/n$ is not an integer for $n \neq \pm 1$. (Note that when learning about new mathematical concepts, it is often very helpful to find such non-examples as well as examples of the concept itself).

A vector space (defined "over a field") is a set containing points, which we can add together and multiply by scalars (i.e. members of the field). This generalises $\mathbb{R}^2$, familiar Euclidean space of dimension *n*, in which we can add coordinates together, and multiply (stretch) vectors by real numbers.

"Projective space", denoted $\mathbb{P}^n$, is defined with respect to a field. It is perhaps best understood as a "quotient space", but can also be thought of as "the set of all lines through the origin" in a vector space (over a field). As a quotient space, for some $n+1$-dimensional vector space V, projective space $\mathbb{P}^n$ is the set of all points in V equated up to scalar multiples, with the origin removed. Note that this is the same as the set of lines through the origin. This gives rise to the notion of "homogeneous coordinates" (also called "projective coordinates"), which are the standard coordinate system for describing points in projective space. Many of our models represent surfaces in projective space, and are defined using homogeneous coordinates.

A relatively easy projective space to visualise is the projective plane $\mathbb{P}^2$, which can be attained by taking all points on a sphere, and "gluing" antipodal points together.

Dimension and Degree

An algebraic surface may also be thought of as an algebraic variety of dimension two. The notion of dimension of an algebraic variety generalises the notions of dimension for the more familiar Euclidean space or a vector space. The dimension of a given variety can be defined as the greatest length *d* of a chain of distinct sub-varieties $X_0 \subset X_1 \subset \cdots \subset X_d$ all contained within each other. For example: a point, line, and plane can form a maximal chain with $l = 2$, so a plane has dimension two, and is hence a (degenerate) algebraic surface. A complex cubic surface is an algebraic surface in $\mathbb{C}$ over the field $\mathbb{C}$ of complex numbers. Recall the definition of the solutions to polynomial equations are considered over the complex numbers rather than the reals, and the polynomials involved can have complex coefficients).

An important property of an algebraic surface is its "degree" (also known as "order"), which may be thought of as the "total degree" of the polynomial defining it. Total degree refers to the sum of the exponents of variables. For example, the polynomial defined by $f(x, y, z) = x^3y - 2y^2z + z^3$ has total degree four. For the more general case of an algebraic variety, the notion of degree is more complicated. It is defined using intersection theory. We have lots of models of algebraic surfaces of order 2, 3, and 4, respectively named quadric, cubic, and quartic surfaces. There are also a few surfaces of higher degree.

Cubic Curve

A cubic curve is an Algebraic Curve of degree 3. An algebraic curve over a Field K is an equation f(X, Y) = 0, where f(X, Y) is a Polynomial in X and Y with Coefficients in K, and the degree of f is the Maximum degree of each of its terms (Monomials).

Newton showed that all cubics can be generated by the projection of the five divergent cubic parabolas. Newton's classification of cubic curves appeared in the chapter Curves in Lexicon Technicum by John Harris published in London in 1710. Newton also classified all cubics into 72 types, missing six of them. In addition, he showed that any cubic can be obtained by a suitable projection of the Elliptic Curve:

$$y^2 = ax^3 + bx^2 + cx + d,$$

Where, the projection is a Birational Transformation, and the general cubic can also be written as:

$$y^2 = x^3 + ax + b.$$

Newton's first class is equations of the form:

$$xy^2 + ey = ax^3 + bx^2 + cx + d.$$

This is the hardest case and includes the Serpentine Curve as one of the subcases. The third class was:

$$ay^2 = x(x^2 - 2bx + c),$$

which is called Newton's Diverging Parabolas. Newton's 66th curve was the Trident of Newton. Newton's classification of cubics was criticized by Euler because it lacked generality. Plücker later gave a more detailed classification with 219 types.

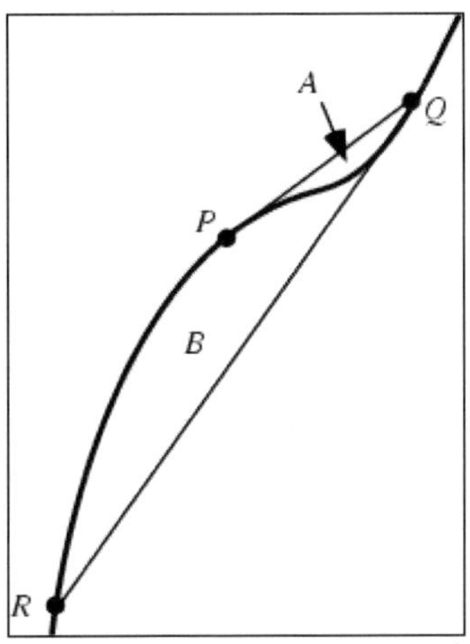

Pick a point P, and draw the tangent to the curve at P. Call the point where this tangent intersects the curve Q. Draw another tangent and call the point of intersection with the curve R. Every curve of third degree has the property that, with the areas in the above labeled figure,

B=16A

Elliptic Curve

Informally, an elliptic curve is a type of cubic curve whose solutions are confined to a region of space that is topologically equivalent to a torus. The Weierstrass elliptic function $P(z; g_2, g_3)$ describes how to get from this torus to the algebraic form of an elliptic curve.

Formally, an elliptic curve over a field K is a nonsingular cubic curve in two variables, f(X, Y) = 0, with a K-rational point (which may be a point at infinity). The field K is usually taken to be the complex numbers $\mathbb{C}$, reals $\mathbb{R}$, rationals $\mathbb{Q}$, algebraic extensions of $\mathbb{Q}$ p-adic numbers $\mathbb{Q}_p$, or a finite field.

By an appropriate change of variables, a general elliptic curve over a field with field characteristic ≠2,3, a general cubic curve:

$$Ax^3 + Bx^2y + Cxy^2 + Dy^3 + Ex^2 + Fxy + Gy^2 + Hx + Iy + J = 0$$

Where, A, B, ..., are elements of K, can be written in the form:

$$y^2 = x^3 + ax + b,$$

where, the right side of $y^2 = x^3 + ax + b$, has no repeated factors. Any elliptic curve not of characteristic $y^2 = x^3 + ax + b$, or Legendre normal form:

$$y^2 = x(x-1)(x-\lambda)$$

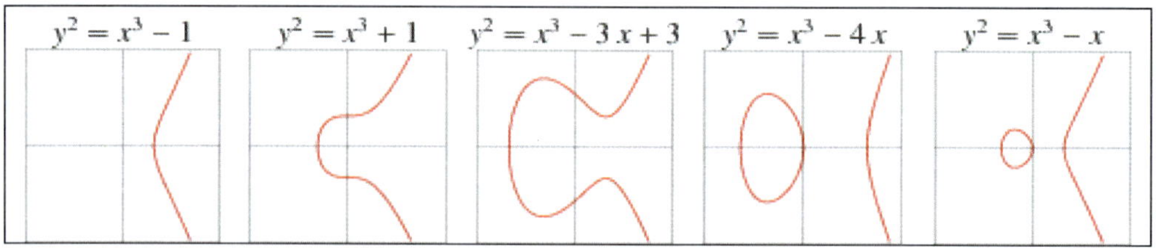

Elliptic curves are illustrated above for various values of a and b.

If K has field characteristic three, then the best that can be done is to transform the curve into:

$$y^2 = x^3 + ax^2 + bx + c$$

(the x^2 term cannot be eliminated). If K has field characteristic two, then the situation is even worse. A general form into which an elliptic curve over any K can be transformed is called the Weierstrass form, and is given by:

$$y^2 + ay = x^3 + bx^2 + cxy + dx + e,$$

where, a, b, c, d, and e are elements of K. Luckily, $\mathbb{Q}$, $\mathbb{R}$, and $\mathbb{C}$ all have field characteristic zero.

An elliptic curve of the form $y^2 = x^3 + n$ for n an integer is known as a Mordell curve.

Whereas conic sections can be parameterized by the rational functions, elliptic curves cannot. The simplest parameterization functions are elliptic functions. Abelian varieties can be viewed as generalizations of elliptic curves.

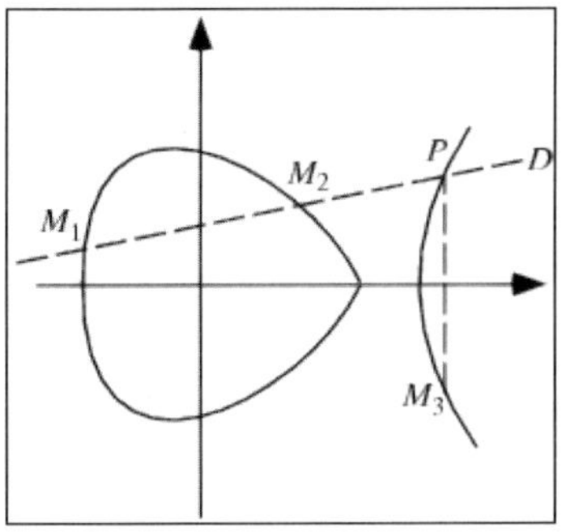

If the underlying field of an elliptic curve is algebraically closed, then a straight line cuts an elliptic curve at three points (counting multiple roots at points of tangency). If two are known, it is possible to compute the third. If two of the intersection points are K-rational, then so is the third. Mazur and Tate proved that there is no elliptic curve over $\mathbb{Q}$ having a rational point of order 13.

Let (x_1, y_1) and (x_2, y_2) be two points on an elliptic curve E with elliptic discriminant:

$$\Delta_E = -16(4a^3 + 27b^2)$$

Satisfying:

$$\Delta_E \neq 0.$$

A related quantity known as the j-invariant of E is defined as:

$$j(E) = \frac{2^8 3^3 a^3}{4a^3 + 27b^2}$$

Now define:

$$\lambda = \begin{cases} \dfrac{y_1 - y_2}{x_1 - x_2} \; for \; x_1 \neq x_2 \\ \dfrac{3x_1^2 + a}{2y_1} \; for \; x_1 = x_2 \end{cases}$$

Then the coordinates of the third point are:

$$x_3 = \lambda^2 - x_1 - x_2$$
$$y_3 = \lambda(x_3 - x_1) + y_1.$$

For elliptic curves over Q, Mordell proved that there are a finite number of integral solutions. The Mordell-Weil theorem says that the group of rational points of an elliptic curve over Q is finitely generated. Let the roots of y^2 be r_1, r_2, and r_3. The discriminant is then:

$$\Delta = k(r_1 - r_2)^2 (r_1 - r_3)^2 (r_2 - r_3)^2.$$

The amazing Taniyama-Shimura conjecture states that all rational elliptic curves are also modular. This fact is far from obvious, and despite the fact that the conjecture was proposed in 1955, it was not even partially proved until 1995. Even so, Wiles' proof for the semistable case surprised most mathematicians, who had believed the conjecture unassailable. As a side benefit, Wiles' proof of the Taniyama-Shimura conjecture also laid to rest the famous and thorny problem which had baffled mathematicians for hundreds of years, Fermat's last theorem.

Curves with small j-conductors are listed in Swinnerton-Dyer and Cremona. Methods for computing integral points (points with integral coordinates) are given in Gebel et al. and Stroeker and Tzanakis. The Schoof-Elkies-Atkin algorithm can be used to determine the order of an elliptic curve E/F_p over the finite field F_p.

Algebraic Variety

An algebraic variety is a generalization to n dimensions of algebraic curves. More technically, an algebraic variety is a reduced scheme of finite type over a field K. An algebraic variety V in $\mathbb{R}^n$ $(or\, \mathbb{C}^n)$ is defined as the set of points satisfying a system of polynomial equations $f_i(x_1,...,x_n)=0$ for i=1, 2, ... According to the Hilbert basis theorem, a finite number of equations suffice.

A variety is the set of common zeros to a collection of polynomials. In classical algebraic geometry, the polynomials have complex numbers for coefficients. Because of the fundamental theorem of algebra, such polynomials always have zeros. For example,

$$\{(x,y,z): x^2 + y^2 - z^2 = 0\}$$

is the cone, and:

$$\{(x,y,z): x^2 + y^2 - z^2 = 0,\ ax + by + cz = 0\}$$

is a conic section, which is a subvariety of the cone.

Actually, the cone and the conic section are examples of affine varieties because they are in affine space. A general variety is comprised of affine varieties glued together, like the coordinate charts of a manifold. The field of coefficients can be any algebraically closed field. When a variety is embedded in projective space, it is a projective algebraic variety. Also, an intrinsic variety can be thought of as an abstract object, like a manifold, independent of any particular embedding. A scheme is a generalization of a variety, which includes the possibility of replacing $\mathbb{C}\,[x,y,z]$ by any commutative ring with a unit. A further generalization is a moduli space stack.

Affine Variety

An afine variety V is an algebraic variety contained in affine space. For example,

$$\{(x,y,z): x^2 + y^2 - z^2 = 0\}$$

is the cone, and:

$$\{(x,y,z): x^2 + y^2 - z^2 = 0,\ ax + by + cz = 0\}$$

is a conic section, which is a subvariety of the cone. The cone can be writtenen V(x^2+y^2-z^2) to indicate that it is the variety corresponding to x^2+y^2-z^2 = 0. Naturally, many other polynomials $I(C) = x^2 + y^2 - z^2$ vanish on V(x^2+y^2-z^2), in fact all polynomials in $I(C) = x^2 + y^2 - z^2$. The set $I(C)$ is an ideal in the polynomial ring $\mathbb{C}\ [x, y, z]$. Note also, that the ideal of polynomials vanishing on the conic section is the ideal generated by x^2+y^2-z^2, and ax+by+cz.

A morphism between two affine varieties is given by polynomial coordinate functions. For example, the mapis a morphism from X=V(x^2+y^2+z^2) to Y=V(x+y+z). Two affine varieties are isomorphic if there is a morphism which has an inverse morphism. For example, the affine variety V(x^2+y^2+z^2) is isomorphic to the cone V(x^2+y^2+z^2) via the coordinate change $\phi(x, y, z) = (x^2, y^2, iz)$.

Many polynomials f may be factored, for $V(f) = V(x+iy) \cup V(x-iy)$ and then $V(f) = V(x+iy) \cup V(x-iy)$. Consequently, only irreducible polynomials, and more generally only prime ideals p are used in the definition of a variety. An affine variety V is the set of common zeros of a collection of polynomials p_1,.......p_k i.e.,

$$V = \{x = (x_1, \ldots\ldots, x_n) : p_1(x) = \ldots\ldots = p_k(x) = 0\}$$

as long as the ideal $I = \langle p_1, \ldots, p_k \rangle$ is a prime ideal. More classically, an affine variety is defined by any set of polynomials, i.e., what is now called an algebraic set. Most points in V will have dimension n-k, but V may have singular points like the origin in the cone.

When V is one-dimensional generically (at almost all points), which typically occurs when k=n-1, then V is called a curve. When V is two-dimensional, it is called a surface. In the case of CW-complex affine space, a curve is a Riemann surface, possibly with some singularities.

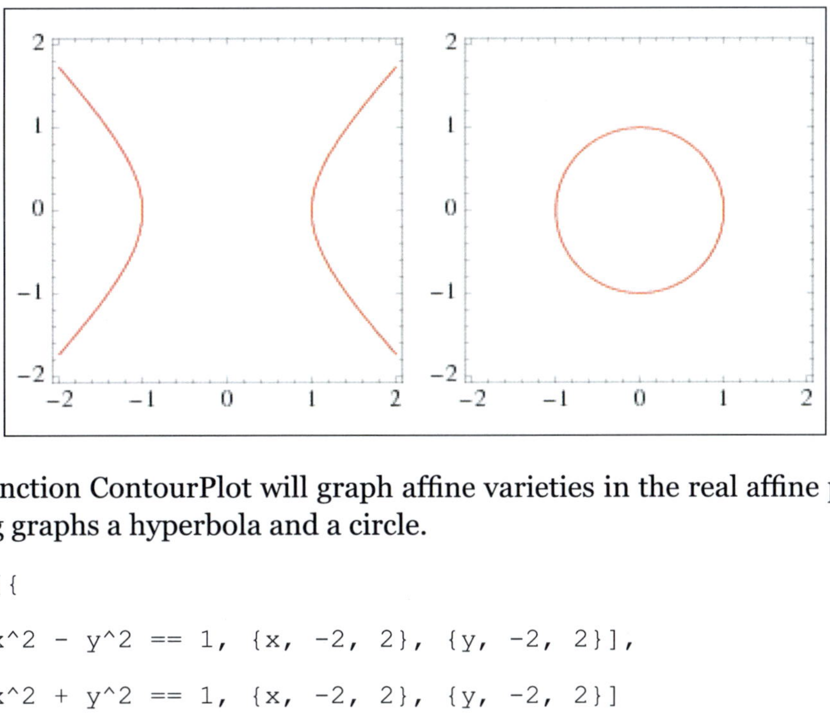

The Language function ContourPlot will graph affine varieties in the real affine plane. For example, the following graphs a hyperbola and a circle.

```
GraphicsGrid[{{

 ContourPlot[x^2 - y^2 == 1, {x, -2, 2}, {y, -2, 2}],

 ContourPlot[x^2 + y^2 == 1, {x, -2, 2}, {y, -2, 2}]

}}]
```

Pascal's Theorem

Pascals Theorem states if we draw a hexagon inscribed in a conic section then the three pairs of opposite sides of the hexagon intersect at three points which lie on a straight line.

For example, as in the following figure we have a hexagon inscribed in a circle and the intersection points of the three pairs of the opposite sides of the hexagon {12, 45}, {23, 56}, {34, 61} are collinear.

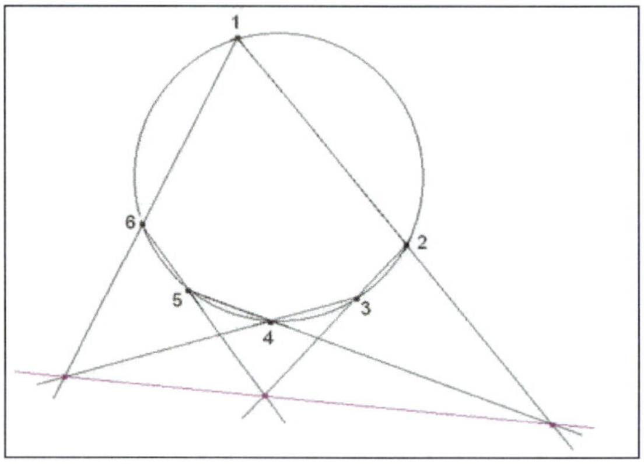

There is a useful tool to prove the collinearity of points - the Menelaus' Theorem - which states as follows:

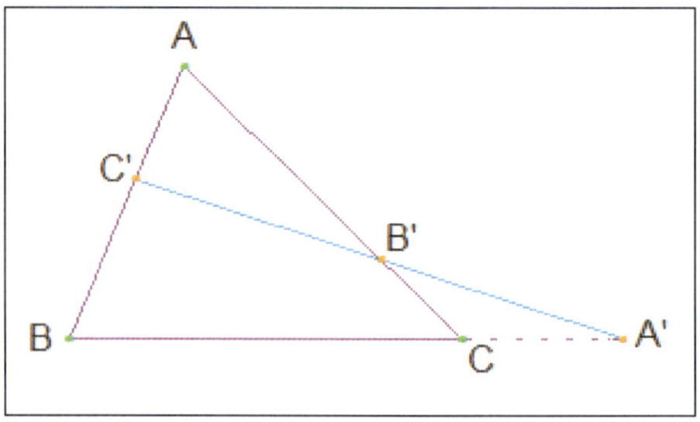

Menelaus' Theorem: Given a triangle ABC and three points A′, B′, C′ lying on the three lines BC, CA, AB, respectively. Then the three points A′, B′, C′ are collinear if and only if:

$$\frac{\overrightarrow{A'B}}{\overrightarrow{A'C}} \times \frac{\overrightarrow{B'C}}{\overrightarrow{B'A}} \times \frac{\overrightarrow{C'A}}{\overrightarrow{C'B}} = 1$$

When we want to use Menelaus' theorem to prove three certain points lying on a same straight line, we need to specify a triangle such that these three points belong to the three sides of the triangle.

In the following figure, for the Pascal's hexagon $P_1P_2P_3P_4P_5P_6$, to show that the three intersection points M_1, M_2, M_3 are collinear, we need to find a triangle whose three sides contain M_1, M_2, M_3. Since, M_1, M_2, M_3 lie on the sides of the hexagon, there are two natural choices for the triangle: either ABC, or XYZ.

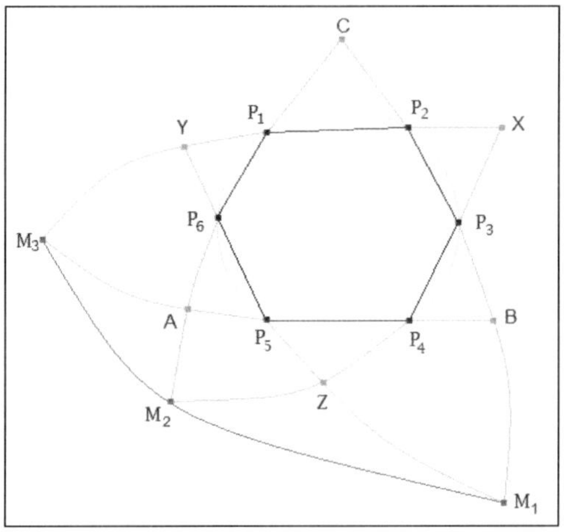

Suppose we choose the triangle ABC, then we can see that the point M_1 lying on the line BC, the point M_2 lying on the line CA and the point M3 lying on the line AB. To use Menelaus' theorem, we need to show:

$$\frac{\overrightarrow{M_1B}}{\overrightarrow{M_1C}} \times \frac{\overrightarrow{M_2C}}{\overrightarrow{M_2A}} \times \frac{\overrightarrow{M_3A}}{\overrightarrow{M_3B}} = 1$$

To calculate these quotients, we will apply Menelaus' theorem for the following triples of collinear points:

$$\{M_1, P_5, P_6\}, \{M_2, P_3, P_4\}, \{M_3, P_1, P_2\}.$$

A Proof of Pascal's Theorem

We will apply Menelaus' theorem for the triangle ABC.

Since, M_1, M_2, M_3 lie on the three sides of the triangle ABC, to prove that they are collinear, we need to show:

$$\frac{\overrightarrow{M_1B}}{\overrightarrow{M_1C}} \times \frac{\overrightarrow{M_2C}}{\overrightarrow{M_2A}} \times \frac{\overrightarrow{M_3A}}{\overrightarrow{M_3B}} = 1$$

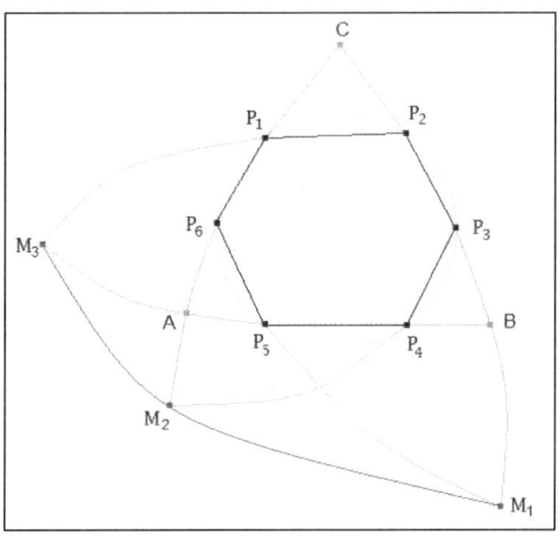

Indeed, apply Menelaus' theorem for the triangle ABC with the following triples of collinear points:

$$\{M_1, P_6, P_5\},\ \{M_2, P_4, P_3\},\ \{M_3, P_2, P_1\},$$

we have

$$\frac{\overrightarrow{M_1B}}{\overrightarrow{M_1C}} \times \frac{\overrightarrow{P_6C}}{\overrightarrow{P_6A}} \times \frac{\overrightarrow{P_5A}}{\overrightarrow{P_5B}} = \frac{\overrightarrow{M_2C}}{\overrightarrow{M_2A}} \times \frac{\overrightarrow{P_4A}}{\overrightarrow{P_4B}} \times \frac{\overrightarrow{P_3B}}{\overrightarrow{P_3C}} = \frac{\overrightarrow{M_3A}}{\overrightarrow{M_3B}} \times \frac{\overrightarrow{P_2B}}{\overrightarrow{P_2C}} \times \frac{\overrightarrow{P_1C}}{\overrightarrow{P_1A}} = 1$$

Thus,

$$\frac{\overrightarrow{M_1B}}{\overrightarrow{M_1C}} = \frac{\overrightarrow{P_6A}}{\overrightarrow{P_6C}} \times \frac{\overrightarrow{P_5B}}{\overrightarrow{P_5A}}, \frac{\overrightarrow{M_2C}}{\overrightarrow{M_2A}} = \frac{\overrightarrow{P_4B}}{\overrightarrow{P_4A}} \times \frac{\overrightarrow{P_3C}}{\overrightarrow{P_3B}}, \frac{\overrightarrow{M_3A}}{\overrightarrow{M_3B}} = \frac{\overrightarrow{P_2C}}{\overrightarrow{P_2B}} \times \frac{\overrightarrow{P_1A}}{\overrightarrow{P_1C}}.$$

It follows that:

$$\frac{\overrightarrow{M_1B}}{\overrightarrow{M_1C}} \times \frac{\overrightarrow{M_2C}}{\overrightarrow{M_2A}} \times \frac{\overrightarrow{M_3A}}{\overrightarrow{M_3B}} = \frac{\overrightarrow{P_6A}}{\overrightarrow{P_6C}} \times \frac{\overrightarrow{P_5B}}{\overrightarrow{P_5A}} \times \frac{\overrightarrow{P_4B}}{\overrightarrow{P_4A}} \times \frac{\overrightarrow{P_3C}}{\overrightarrow{P_3B}} \times \frac{\overrightarrow{P_2C}}{\overrightarrow{P_2B}} \times \frac{\overrightarrow{P_1A}}{\overrightarrow{P_1C}}$$

$$\frac{\overrightarrow{AP_1}\,\overrightarrow{AP_6}}{\overrightarrow{AP_4}\,\overrightarrow{AP_5}} \times \frac{\overrightarrow{BP_4}\,\overrightarrow{BP_5}}{\overrightarrow{BP_1}\,\overrightarrow{BP_3}} \times \frac{\overrightarrow{CP_2}\,\overrightarrow{CP_3}}{\overrightarrow{CP_1}\,\overrightarrow{CP_6}} = 1.$$

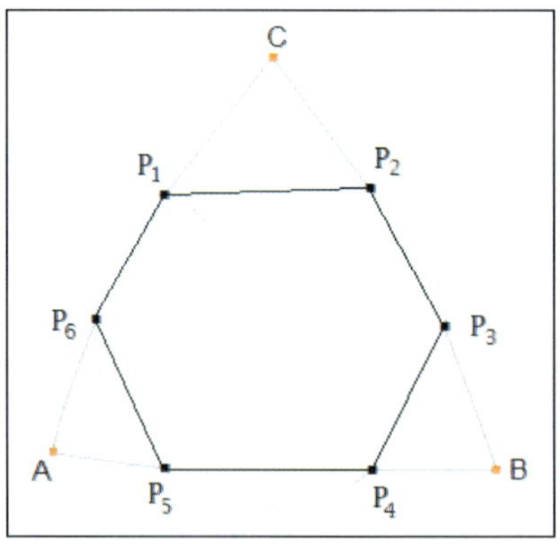

Power of a point: $\overrightarrow{AP_1}\,\overrightarrow{AP_6} = \overrightarrow{AP_4}\,\overrightarrow{AP_5};\ \overrightarrow{BP_4}\,\overrightarrow{BP_5} = \overrightarrow{BP_2}\,\overrightarrow{BP_3}; \overrightarrow{CP_2}\,\overrightarrow{CP_3} = \overrightarrow{CP_1}\,\overrightarrow{CP_6}$

The last equality is from the power of a point property when we apply the power of the points A, B, C to the circumcircle of the hexagon. Thus, we have completed the proof of Pascal's theorem.

Riemann-Roch Theorem

In algebraic geometry the Riemann-Roch theorem states that if C is a smooth algebraic curve, and $\mathcal{L}$ is an invertible sheaf on C then the following properties hold:

The Euler characteristic of $\mathcal{L}$ is given by:

$$h^0(\mathcal{L}) - h^1(\mathcal{L}) = \deg(\mathcal{L}) - (g - 1)$$

There is a canonical isomorphism,

$$H^0(L)(K_C \otimes \mathcal{L}^\vee) \cong H^1(\mathcal{L})$$

Some Examples and Applications

The examples we give arise from considering complete linear systems on curves.

Any curve C of genus 0 is isomorphic to the projective line: Indeed if p is a point on the curve then $h^o(p) - 0 = 1 - (0 - 1) = 2$; hence the map $C \to \mathbb{P}H^0(O_C(p))$ is a degree 1 map, or an isomorphism.

Any curve of genus 1 is a double cover of a projective line: Indeed if p is a point on the curve then $h^o(2p) - 0 = 2 - (1 - 1) = 2$; hence the map $C \to \mathbb{P}H^0(O_C(2p))$ is a degree 2 map.

Any curve of genus 2 is a double cover of a projective line: Indeed the degree of the canonical class K_C is $2g - 2$ and therefore $h^o(K_C) - h^o(O_C) = 2 - (2 - 1) = 1$; since $h^o(O_C) = 1$ the map $C \to \mathbb{P}H^0(K_C)$ is a degree 2 map.

The Riemann-Hurwitz formula.

Geometric Riemann-Roch

From the statement of the theorem one sees that an effective divisor D of degree d on a curve C satisfies $h^o(D) > d - (g - 1)$ if and only if there is an effective divisor D' such that $D + D' \sim K_C$ in $Pic(C)$. In this case there is a natural isomorphism:

$$\{[H] \in | K_C |, H \cdot C = D'\} \cong \mathbb{P}H^0(D),$$

where, we identify C with it's image in the dual canonical system $| K_C |^*$.

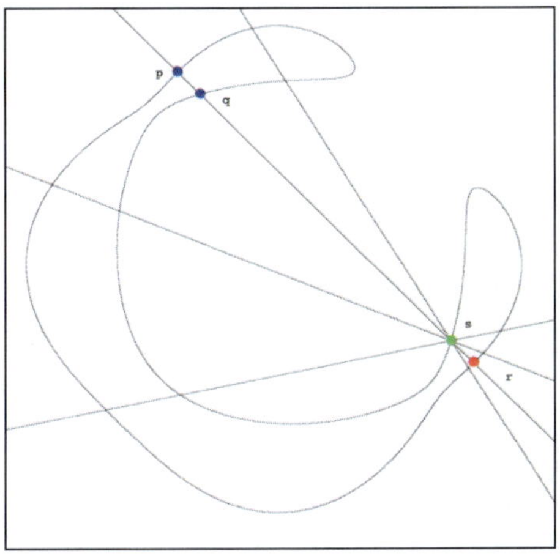

Some linear systems on a smooth cannonicaly embedded genus-3 curve.

As an example we consider effective divisors of degrees 2,3 on a non hyperelliptic curve C of genus 3. The degree of the canonical class is 2genus(c) – 2 = 4, whereas $h^o(K_C)$ = 2genus(C) – 2 – (genus(C) – 1) + $h^o(0) = g$. Hence the canonical image of C is a smooth plane quartic. We now idenitfy C with it's image in the dual canonical system. Let p,q be two points on C then there are exactly two points r,s such that $C \cap \overline{pq} = \{p,q,r,s\}$, where we intersect with multiplicities, and if $p = q$ we consider the tangent line T_pC instead of the line $\overline{pq}$. Hence there is a natural isomorphism between $\mathbb{P}h^0(O_C(p+q))$ and the unique point in $|K_C|$ representing the line $\overline{pq}$. There is also a natural isomorphism between $\mathbb{P}(O_C(p+q+r))$ and the points in $|K_C|$ representing lines through the points s.

Proofs

In 1857 Riemann proved the inequality $h^0(O_C(D)) \leq \deg(D)-(g-1)$; the proof used function theoretic methods on a gluing diagram of a 4g-gon. In 1865 Riemann's student Roch identified the difference in the inequality as the dimension of the cohomology group of differentials with prescribed zeros given by $H^0(O_C(K-D))$. In the second half of the 19th century, and in the beginning of the 20th century, a huge effort went into making the proof more geometric, and later more algebraic. (i.e. to get a proof over function fields which is independent of the geometric model). The problem in "geometrizing" the theorem is that in the 19th century, curves would always arise as embedded curves in some ambient projective space, and one had to find birationaly invariant definitions for the cohomology groups involved in the theorem. The problem in "algebraizing" the theorem is that one has to define differentials over function fields. In subsequent years the analytic, geometric and algebraic proofs have all been considerably simplified. The current "standard" proof is the sheaf theoretic one: using modern tools, the theorem is an immediate consequence of Serre's duality, and the fact that if D, D' are divisors on C then $\chi(O_C(D+D')) = \chi(O_C(D)) + \chi(O_C(D'))$.

Generalizations

The generalizations of the Riemann-Roch theorem come in two flavors: One direction views the Riemann-Roch theorem as a tool to study any linear system on a any curve. Clifford's theorem gives a better bound on the dimension of special linear systems on curve. If the assumption on the curve is relaxed to be a generic curve, then the Brill-Noether Theorem and the Petri Theorem give good descriptions of the geometry of the linear system.

Another direction of generalization, with more far-reaching consequences, is to view Riemann-Roch as a tool to compute the Euler characteristic of a vector bundle on a Variety. The first generalizations in this direction go back to the beginning of the 20th century with Riemann-Roch for surfaces and Noether's formula on surfaces. The next step, taken during the 1960s, is the Hirzebruch-Riemann-Roch theorem, which analyze the Euler characteristic of the canonical bundle of an arbitrary. The final step in the algebro-geometric setting is the Grothendieck-Riemann-Roch theorem, which analyzes the behaviour of the Euler characteristic of vector bundles under pullbacks; e.g. the Riemann-Roch theorem can be deduced from the Grothendieck-Riemann Roch theorem by projecting a curve to a point. In the analytic setting Grothendieck Riemann Roch had one more generalization: the Atiyah-Singer index theorem.

References

- Algebraic-geometry, theory: maths.ox.ac.uk, Retrieved 27 August, 2019
- EllipticCurve: mathworld.wolfram.com, Retrieved 05 March, 2019
- AlgebraicVariety: mathworld.wolfram.com, Retrieved 13 July, 2019
- Riemann-Roch-theorem: m.tau.ac.il, Retrieved 17 March, 2019
- AffineVariety: mathworld.wolfram.com, Retrieved 12 February, 2019

6

Analytic Geometry

Analytic geometry is a branch of algebra that deals with the modeling of geometric objects such as points, lines and circles by using a Cartesian system. Conic sections and tangents and normal are also studied under it. This chapter sheds light on analytic geometry for a thorough understanding of the subject.

Analytic geometry, also called coordinate geometry is a mathematical subject in which algebraic symbolism and methods are used to represent and solve problems in geometry. The importance of analytic geometry is that it establishes a correspondence between geometric curves and algebraic equations. This correspondence makes it possible to reformulate problems in geometry as equivalent problems in algebra, and vice versa; the methods of either subject can then be used to solve problems in the other. For example, computers create animations for display in games and films by manipulating algebraic equations.

Elementary Analytic Geometry

Apollonius of Perga, known by his contemporaries as the "Great Geometer," fore-shadowed the development of analytic geometry by more than 1,800 years with his book Conics. He defined a conic as the intersection of a cone and a plane. Using Euclid's results on similar triangles and on secants of circles, he found a relation satisfied by the distances from any point P of a conic to two perpendicular lines, the major axis of the conic and the tangent at an endpoint of the axis. These distances correspond to coordinates of P, and the relation between these coordinates corresponds to a quadratic equation of the conic. Apollonius used this relation to deduce fundamental properties of conics.

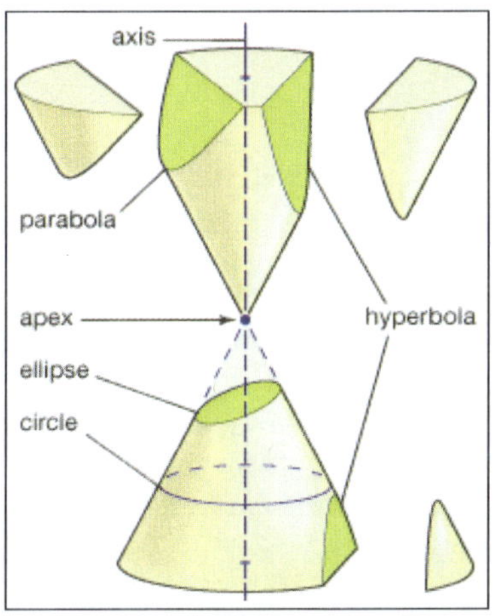

Conic sections: The conic sections result from intersecting a plane with a double cone, as shown in the figure. There are three distinct families of conic sections: The ellipse (including the circle); the parabola (with one branch); and the hyperbola (with two branches).

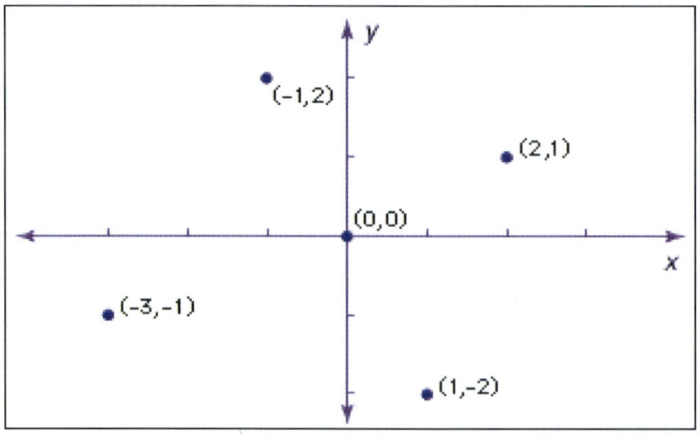

Cartesian coordinates: Several points are labeled in a two-dimensional graph, known as the Cartesian plane. Note that each point has two coordinates, the first number (x value) indicates its distance from the y-axis—positive values to the right and negative values to the left—and the second number (y value) gives its distance from the x-axis—positive values upward and negative values downward.

Descartes and Fermat independently founded analytic geometry in the 1630s by adapting Viète's algebra to the study of geometric loci. They moved decisively beyond Viète by using letters to represent distances that are variable instead of fixed. Descartes used equations to study curves defined geometrically, and he stressed the need to consider general algebraic curves—graphs of polynomial equations in x and y of all degrees. He demonstrated his method on a classical problem: finding all points P such that the product of the distances from P to certain lines equals the product of the distances to other lines.

Fermat emphasized that any relation between x and y coordinates determines a curve. Using this idea, he recast Apollonius's arguments in algebraic terms and restored lost work. Fermat indicated that any quadratic equation in x and y can be put into the standard form of one of the conic sections.

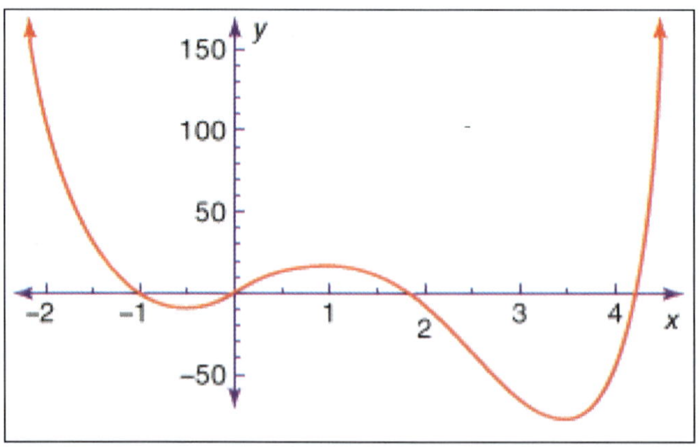

Polynomial graph: The figure shows part of the graph of the polynomial equation $y = 3x^4 - 16x^3 + 6x^2 + 24x + 1$. Note that the same scale need not be used for the x- and y-axis.

Fermat did not publish his work, and Descartes deliberately made his hard to read in order to discourage "dabblers." Their ideas gained general acceptance only through the efforts of other mathematicians in the latter half of the 17th century. In particular, the Dutch mathematician Frans van Schooten translated Descartes's writings from French to Latin. He added vital explanatory material, as did the French lawyer Florimond de Beaune, and the Dutch mathematician Johan de Witt. In England, the mathematician John Wallis popularized analytic geometry, using equations to define conics and derive their properties. He used negative coordinates freely, although it was Isaac Newton who unequivocally used two (oblique) axes to divide the plane into four quadrants, as shown in the figure.

Analytic geometry had its greatest impact on mathematics via calculus. Without access to the power of analytic geometry, classical Greek mathematicians such as Archimedes (c. 285–212/211 BC) solved special cases of the basic problems of calculus: finding tangents and extreme points (differential calculus) and arc lengths, areas, and volumes (integral calculus). Renaissance mathematicians were led back to these problems by the needs of astronomy, optics, navigation, warfare, and commerce. They naturally sought to use the power of algebra to define and analyze a growing range of curves.

Fermat developed an algebraic algorithm for finding the tangent to an algebraic curve at a point by finding a line that has a double intersection with the curve at the point—in essence, inventing differential calculus. Descartes introduced a similar but more complicated algorithm using a circle. Fermat computed areas under the curves y = axk for all rational numbers k ≠ −1 by summing areas of inscribed and circumscribed rectangles. For the rest of the 17th century, the groundwork for calculus was continued by many mathematicians, including the Frenchman Gilles Personne de Roberval, the Italian Bonaventura Cavalieri, and the Britons James Gregory, John Wallis, and Isaac Barrow.

Newton and the German Gottfried Leibniz revolutionized mathematics at the end of the 17th century by independently demonstrating the power of calculus. Both men used coordinates to develop notations that expressed the ideas of calculus in full generality and led naturally to differentiation rules and the fundamental theorem of calculus (connecting differential and integral calculus).

Newton demonstrated the importance of analytic methods in geometry, apart from their role in calculus, when he asserted that any cubic—or, algebraic curve of degree three—has one of four standard equations,

$$xy^2+ey=ax^3+bx^2+cx+d,$$

$$xy=ax^3+bx^2+cx+d,$$

$$y^2=ax^3+bx^2+cx+d,$$

$$y=ax^3+bx^2+cx+d,$$

for suitable coordinate axes. The Scottish mathematician James Stirling proved this assertion in 1717, possibly with Newton's aid. Newton divided cubics into 72 species, a total later corrected to 78.

Newton also showed how to express an algebraic curve near the origin in terms of the fractional power series $y=a_1x^{1/k}+a_2x^{2/k}+\ldots$ for a positive integer k. Mathematicians have since used this technique to study algebraic curves of all degrees.

Coordinates

Coordinates are numbers which determine the position of a point or a shape in a particular space (a map or a graph). Consider the following figure:

	A	B	C	D	E	F
1						
2						
3				X		
4						
5						
6						

The columns of the grid are labeled as A, B, C, D, E, F, etc. On the other hand, the rows are numbered as 1, 2, 3, 4, 5, 6, and so on. You can see that the letter X is located in the box D3 i.e. column D and row 3. Here, D and 3 are the coordinates of this box.

The box has two parts – one is the row and the other is the column. You need to understand that there are several boxes in every row and several boxes in every column. So, when you have both of them, you can find one single box that is the point where the rows and the columns intersect each other.

The Coordinate Plane

In the coordinate geometry, all the points are located on the coordinate plane.

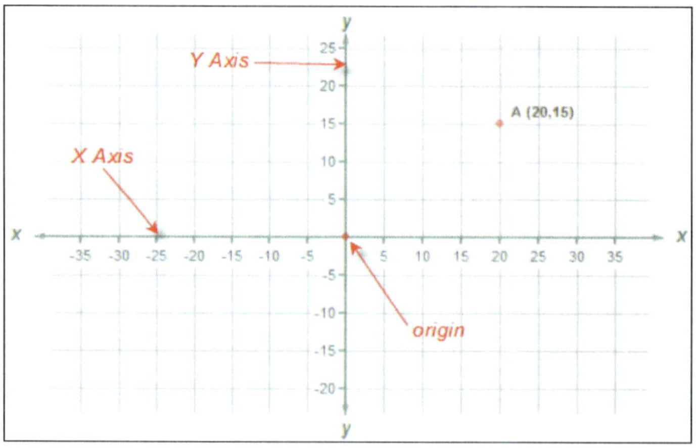

The figure above has two scales – One is the X-axis which is running across the plane and the other one is the y-axis which is at the right angles to the X-axis.

The Concept of Coordinates

- The point of intersection of the x and the y-axis is known as the origin. At this point, both x and y are 0.
- The values on the right-hand side of the x-axis are positive and the values on the left-hand side of the x-axis are negative.

- Similarly, on the y-axis, the values located above the origin are positive and the values located below the origin are negative.
- When you have to locate a point on the plane, it is determined by a set of two numbers. So, first, you have to write about its location on the x-axis followed by its location on the y-axis. Together, the two will determine a single and unique position on the plane.

So, in the figure above, the point A has a value 20 on the x-axis and value 15 on the y-axis. These are also the coordinates of the point A. Often these points are also regarded as the "rectangular coordinates".

Cartesian Coordinate System

In mathematics, the Cartesian coordinate system (or rectangular coordinate system) is used to determine each point uniquely in a plane through two numbers, usually called the *x-coordinate* and the *y-coordinate* of the point. To define the coordinates, two perpendicular directed lines (the *x-axis* or abscissa, and the *y-axis* or ordinate), are specified, as well as the unit length, which is marked off on the two axes. Cartesian coordinate systems are also used in space (where three coordinates are used) and in higher dimensions.

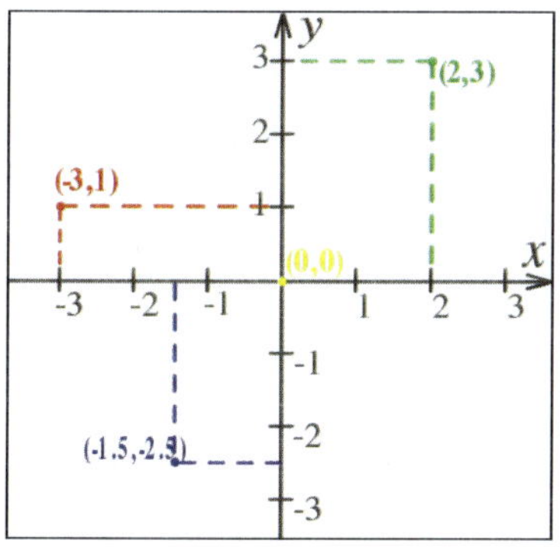

Cartesian coordinate system. Four points are marked: (2, 3) in green, (-3, 1) in red, (-1.5, -2.5) in blue and (0, 0), the origin, in yellow.

Using the Cartesian coordinate system, geometric shapes (such as curves) can be described by algebraic equations, namely equations satisfied by the coordinates of the points lying on the shape. For example, a circle of radius 2 may be described by the equation $x^2 + y^2 = 4$.

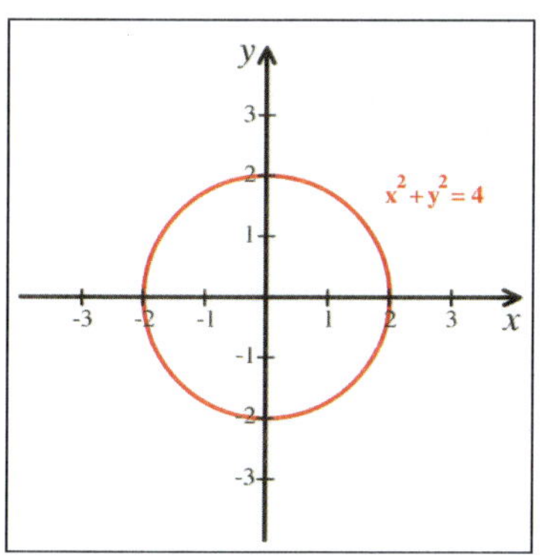

Cartesian coordinate system with the circle of radius 2 centered at the origin marked in red. The equation of the circle is $x^2 + y^2 = 4$.

Two-dimensional Coordinate System

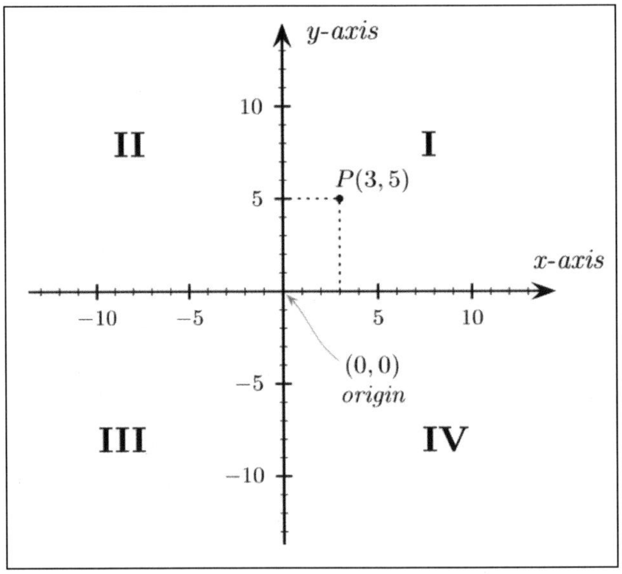

The four quadrants of a Cartesian coordinate system. The arrows on the axes indicate that they extend forever in their respective directions (i.e. infinitely).

A Cartesian coordinate system in two dimensions is commonly defined by two axes, at right angles to each other, forming a plane (an xy-plane). The horizontal axis is normally labeled x, and the vertical axis is normally labeled y. In a three–dimensional coordinate system, another axis, normally labeled z, is added, providing a third dimension of space measurement. The axes are commonly defined as mutually orthogonal to each other (each at a right angle to the other). (Early systems allowed "oblique" axes, that is, axes that did not meet at right angles, and such systems are occasionally used today, although mostly as theoretical exercises). All the points in a Cartesian coordinate system taken together form a so-called Cartesian plane. Equations that use the Cartesian coordinate system are called Cartesian equations.

The point of intersection, where the axes meet, is called the origin normally labeled O. The x and y axes define a plane that is referred to as the xy plane. Given each axis, choose a unit length, and mark off each unit along the axis, forming a grid. To specify a particular point on a two dimensional coordinate system, indicate the x unit first (abscissa), followed by the y unit (ordinate) in the form (x, y), an ordered pair.

The choice of letters comes from a convention, to use the latter part of the alphabet to indicate unknown values. In contrast, the first part of the alphabet was used to designate known values.

An example of a point P on the system is indicated in figure, using the coordinate (3, 5).

The intersection of the two axes creates four regions, called quadrants, indicated by the Roman numerals I (+, +), II (−, +), III (−, −), and IV (+, −). Conventionally, the quadrants are labeled counter-clockwise starting from the upper right ("northeast") quadrant. In the first quadrant, both coordinates are positive, in the second quadrant x-coordinates are negative and y-coordinates positive, in the third quadrant both coordinates are negative and in the fourth quadrant, x-coordinates are positive and y-coordinates negative.

Three-dimensional Coordinate System

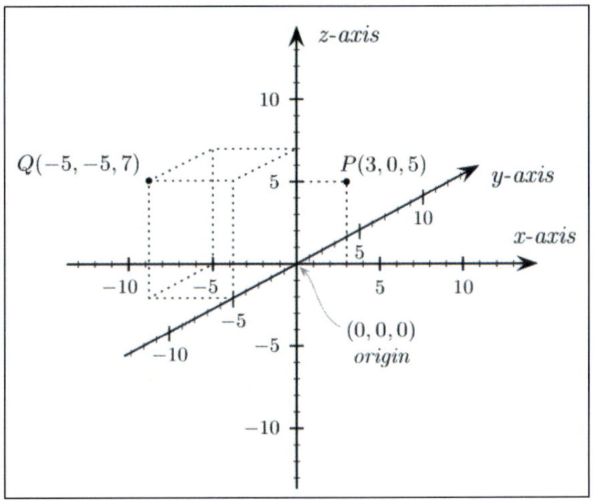

Three dimensional Cartesian coordinate system with y-axis pointing *away* from the observer.

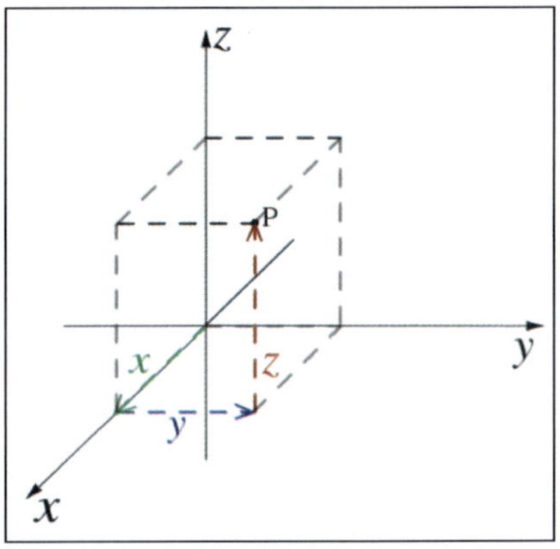

Three dimensional Cartesian coordinate system with the x-axis pointing *towards* the observer.

The three dimensional Cartesian coordinate system provides the three physical dimensions of space—length, width, and height. Figures, show two common ways of representing it.

The three Cartesian axes defining the system are perpendicular to each other. The relevant coordinates are of the form (x, y, z). As an example, figure shows two points plotted in a three-dimensional Cartesian coordinate system: $P(3,0,5)$ and $Q(-5,-5,7)$. The axes are depicted in a "world-coordinates" orientation with the z-axis pointing up.

The x-, y-, and z-coordinates of a point can also be taken as the distances from the yz-plane, xz-plane, and xy-plane respectively. Figure shows the distances of point P from the planes.

The xy-, yz-, and xz-planes divide the three-dimensional space into eight subdivisions known as octants, similar to the quadrants of 2D space. While conventions have been established for the labelling of the four quadrants of the x-y plane, only the first octant of three dimensional space is labelled. It contains all of the points whose x, y, and z coordinates are positive.

The z-coordinate is also called applicate.

Orientation and Handedness

In Two Dimensions

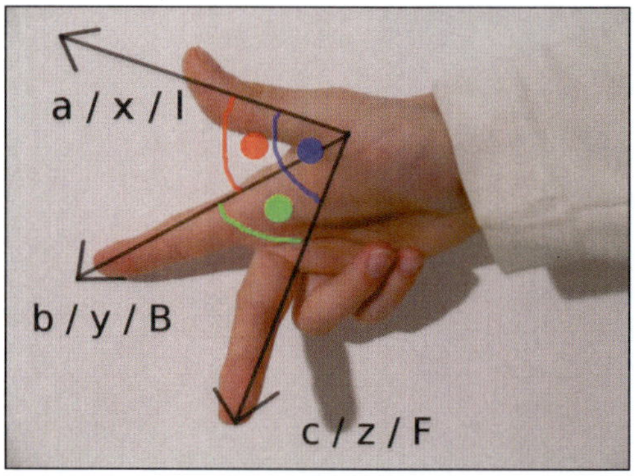

The right hand rule.

Fixing or choosing the x-axis determines the y-axis up to direction. Namely, the y-axis is necessarily the perpendicular to the x-axis through the point marked o on the x-axis. But there is a choice of which of the two half lines on the perpendicular to designate as positive and which as negative. Each of these two choices determines a different orientation (also called *handedness*) of the Cartesian plane.

The usual way of orienting the axes, with the positive x-axis pointing right and the positive y-axis pointing up (and the x-axis being the "first" and the y-axis the "second" axis) is considered the *positive* or *standard* orientation, also called the *right-handed* orientation.

A commonly used mnemonic for defining the positive orientation is the *right hand rule*. Placing a somewhat closed right hand on the plane with the thumb pointing up, the fingers point from the x-axis to the y-axis, in a positively oriented coordinate system.

The other way of orienting the axes is following the *left hand rule*, placing the left hand on the plane with the thumb pointing up.

Regardless of the rule used to orient the axes, rotating the coordinate system will preserve the orientation. Switching the role of x and y will reverse the orientation.

In Three Dimensions

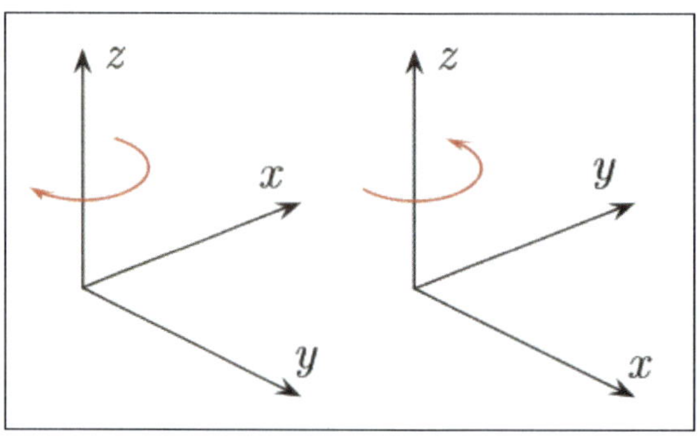

The left-handed orientation is shown on the left, and the right-handed on the right.

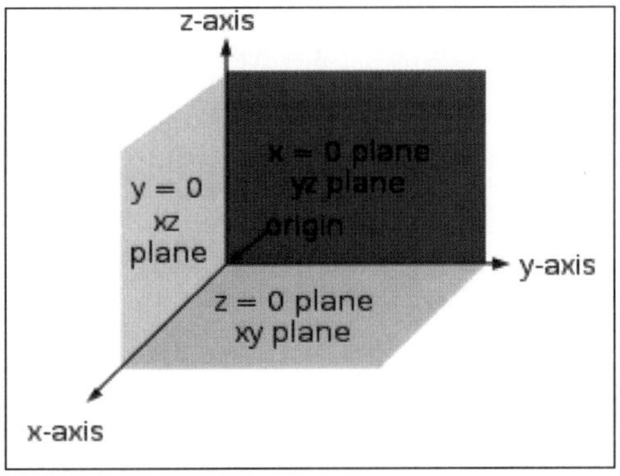

The right-handed Cartesian coordinate system indicating the coordinate planes.

Once the x- and y-axes are specified, they determine the line along which the z-axis should lie, but there are two possible directions on this line. The two possible coordinate systems which result are called "right-handed" and "left-handed." The standard orientation, where the xy-plane is horizontal and the z-axis points up (and the x- and the y-axis form a positively oriented two-dimensional coordinate system in the xy-plane if observed from *above* the xy-plane) is called right-handed or positive.

The name derives from the right-hand rule. If the index finger of the right hand is pointed forward, the middle finger bent inward at a right angle to it, and the thumb placed at a right angle to both, the three fingers indicate the relative directions of the x-, y-, and z-axes in a *right-handed* system. The thumb indicates the x-axis, the index finger the y-axis and the middle finger the z-axis. Conversely, if the same is done with the left hand, a left-handed system results.

Different disciplines use different variations of the coordinate systems. For example, mathematicians typically use a right-handed coordinate system with the y-axis pointing up, while engineers typically use a left-handed coordinate system with the z-axis pointing up. This has the potential to lead to confusion when engineers and mathematicians work on the same project.

Figure is an attempt at depicting a left- and a right-handed coordinate system. Because a three-dimensional object is represented on the two-dimensional screen, distortion and ambiguity result. The axis pointing downward (and to the right) is also meant to point towards the observer, whereas the "middle" axis is meant to point away from the observer. The red circle is parallel to the horizontal xy-plane and indicates rotation from the x-axis to the y-axis (in both cases). Hence the red arrow passes in front of the z-axis.

Figure is another attempt at depicting a right-handed coordinate system. Again, there is an ambiguity caused by projecting the three-dimensional coordinate system into the plane. Many observers see figure as "flipping in and out" between a convex cube and a concave "corner." This corresponds to the two possible orientations of the coordinate system. Seeing the figure as convex gives a left-handed coordinate system. Thus, the "correct" way to view figure is to imagine the x-axis as pointing towards the observer and thus seeing a concave corner.

Polar Coordinate System

In mathematics, the polar coordinate system is a two-dimensional coordinate system in which each point on a plane is determined by a distance from a reference point and an angle from a reference direction. The reference point (analogous to the origin of a Cartesian coordinate system) is called the pole, and the ray from the pole in the reference direction is the polar axis. The distance from the pole is called the radial coordinate, radial distance or simply radius, and the angle is called the angular coordinate, polar angle, or azimuth. The radial coordinate is often denoted by r or ρ, and the angular coordinate by φ, θ, or t. Angles in polar notation are generally expressed in either degrees or radians (2π rad being equal to 360°).

Grégoire de Saint-Vincent and Bonaventura Cavalieri independently introduced the concepts in the mid-seventeenth century, though the actual term polar coordinates has been attributed to Gregorio Fontana in the 18th-century. The initial motivation for the introduction of the polar system was the study of circular and orbital motion.

Polar coordinates are most appropriate in any context where the phenomenon being considered is inherently tied to direction and length from a center point. The Archimedean spiral for example is much simpler to express using polar forms. Moreover, many physical systems—such as those concerned with bodies moving around a central point or with phenomena originating from a central point—are simpler and more intuitive to model using polar coordinates.

The polar coordinate system is extended into three dimensions with two different coordinate systems, the cylindrical and spherical coordinate system. Alexis Clairaut was the first to think of polar coordinates in three dimensions, and Leonhard Euler was the first to actually develop them.

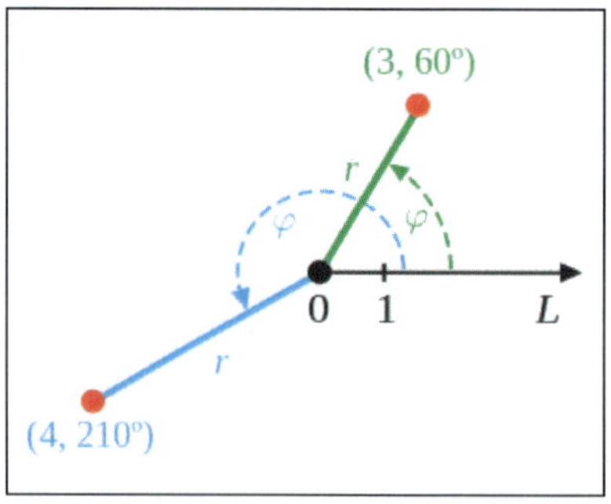

Points in the polar coordinate system with pole O and polar axis L. In green, the point with radial coordinate 3 and angular coordinate 60 degrees or (3, 60°). In blue, the point (4, 210°).

Conventions

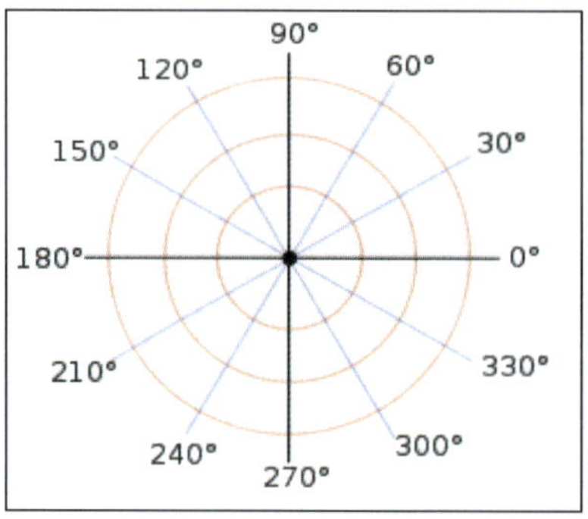

A polar grid with several angles, increasing in ccw orientation and labelled in degrees.

The radial coordinate is often denoted by r or ρ, and the angular coordinate by φ, θ, or t. The angular coordinate is specified as φ by ISO standard 31-11.

Angles in polar notation are generally expressed in either degrees or radians (2π rad being equal to 360°). Degrees are traditionally used in navigation, surveying, and many applied disciplines, while radians are more common in mathematics and mathematical physics.

The angle φ is defined to start at 0° from a *reference direction*, and to increase for rotations in either counterclockwise (ccw) or clockwise (cw) orientation. In mathematics, e.g., the reference direction is usually drawn as a ray from the pole horizontally to the right, and the polar angle increases to positive angles for ccw rotations, whereas in navigation (bearing, heading) the 0°-heading is drawn vertically upwards and the angle increases for cw rotations. The polar angles decrease towards negative values for rotations in the respectively opposite orientations.

Uniqueness of Polar Coordinates

Adding any number of full turns (360°) to the angular coordinate does not change the corresponding direction. Similarly, any polar coordinate is identical to the coordinate with the negative radial component and the opposite direction (adding 180° to the polar angle). Therefore, the same point (r,φ) can be expressed with an infinite number of different polar coordinates $(r,\varphi+n\times 360°)$ and $(-r,\varphi+180°+n\times 360°)=(-r,\varphi+(2n+1)\times 180°)$, where n is an arbitrary integer. Moreover, the pole itself can be expressed as $(0,\varphi)$ for any angle φ.

Where a unique representation is needed for any point besides the pole, it is usual to limit r to positive numbers ($r > 0$) and φ to the interval [0, 360°] or [−180°, 180°] (in radians, $[0, 2\pi]$ or $[-\pi, \pi]$). Another convention, in reverence to the usual codomain of the arctan-function, is to allow for arbitrary nonzero real values of the radial component and restrict the polar angle to [−90°, 90°]. In all cases a unique azimuth for the pole ($r = 0$) must be chosen, e.g., $\varphi = 0$.

Converting between Polar and Cartesian Coordinates

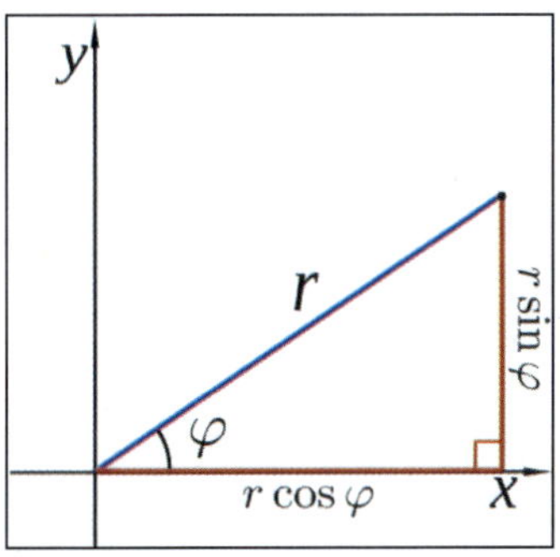

A diagram illustrating the relationship between polar and Cartesian coordinates.

The polar coordinates r and φ can be converted to the Cartesian coordinates x and y by using the trigonometric functions sine and cosine:

$$x = r\cos\varphi,$$
$$y = r\sin\varphi.$$

The Cartesian coordinates x and y can be converted to polar coordinates r and φ with $r \geq 0$ and φ in the interval $(-\pi, \pi]$ by:

$r = \sqrt{x^2 + y^2}$ (as in the Pythagorean theorem or the Euclidean norm), and

$\varphi = \operatorname{atan2}(y, x)$,

where atan2 is a common variation on the arctangent function defined as,

$$\operatorname{atan2}(y,x) = \begin{cases} \arctan\left(\frac{y}{x}\right) & \text{if } x > 0 \\ \arctan\left(\frac{y}{x}\right) + \pi & \text{if } x < 0 \text{ and } y \geq 0 \\ \arctan\left(\frac{y}{x}\right) - \pi & \text{if } x < 0 \text{ and } y < 0 \\ \frac{\pi}{2} & \text{if } x = 0 \text{ and } y > 0 \\ -\frac{\pi}{2} & \text{if } x = 0 \text{ and } y < 0 \\ \text{undefined} & \text{if } x = 0 \text{ and } y = 0. \end{cases}$$

If r is calculated first as above, then this formula for φ may be stated a little more simply using the standard arccosine function:

$$\varphi = \begin{cases} \arccos\left(\frac{x}{r}\right) & \text{if } y \geq 0 \text{ and } r \neq 0 \\ -\arccos\left(\frac{x}{r}\right) & \text{if } y < 0 \\ \text{undefined} & \text{if } r = 0. \end{cases}$$

The value of φ above is the principal value of the complex number function arg applied to $x + iy$. An angle in the range $[0, 2\pi)$ may be obtained by adding 2π to the value in case it is negative (in other words when y is negative).

Polar Equation of a Curve

The equation defining an algebraic curve expressed in polar coordinates is known as a *polar equation*. In many cases, such an equation can simply be specified by defining r as a function of φ. The resulting curve then consists of points of the form $(r(\varphi), \varphi)$ and can be regarded as the graph of the polar function r. Note that, in contrast to Cartesian coordinates, the independent variable φ is the *second* entry in the ordered pair.

Different forms of symmetry can be deduced from the equation of a polar function r. If $r(-\varphi) = r(\varphi)$ the curve will be symmetrical about the horizontal (0°/180°) ray, if $r(\pi - \varphi) = r(\varphi)$ it will be symmetric about the vertical (90°/270°) ray, and if $r(\varphi-\alpha)=r(\varphi)$ it will be rotationally symmetric by α clockwise and counterclockwise about the pole.

Because of the circular nature of the polar coordinate system, many curves can be described by a rather simple polar equation, whereas their Cartesian form is much more intricate. Among the best known of these curves are the polar rose, Archimedean spiral, lemniscate, limaçon, and cardioid.

For the circle, line, and polar rose below, it is understood that there are no restrictions on the domain and range of the curve.

Circle

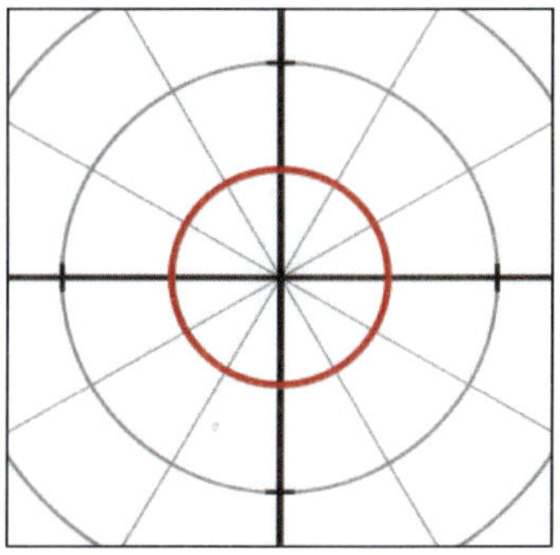

A circle with equation $r(\varphi) = 1$.

The general equation for a circle with a center at (r_0, γ) and radius a is,

$$r^2 - 2rr_0 \cos(\varphi-\gamma) + r_0^2 = a^2$$

This can be simplified in various ways, to conform to more specific cases, such as the equation,

$$r(\varphi) = a$$

for a circle with a center at the pole and radius a.

When r_0 = a, or when the origin lies on the circle, the equation becomes,

$$r = 2a\cos(\varphi-\gamma)$$

In the general case, the equation can be solved for r, giving

$$r = r_0\cos(\varphi-\gamma) + \sqrt{a^2 - r_0^2\sin^2(\varphi-\gamma)},$$

the solution with a minus sign in front of the square root gives the same curve.

Line

Radial lines (those running through the pole) are represented by the equation,

$$\varphi = \gamma,$$

where γ is the angle of elevation of the line; that is, γ = arctan m, where m is the slope of the line in the Cartesian coordinate system. The non-radial line that crosses the radial line $\varphi = \gamma$ perpendicularly at the point (ro, γ) has the equation,

$$r(\varphi) = r_0 \sec(\varphi - \gamma)$$

Otherwise stated (ro, γ) is the point in which the tangent intersects the imaginary circle of radius r_0.

Polar Rose

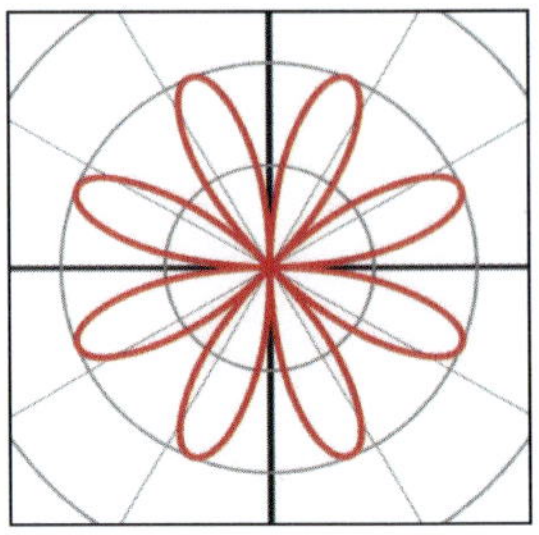

A polar rose with equation $r(\varphi) = 2 \sin 4\varphi$.

A polar rose is a mathematical curve that looks like a petaled flower, and that can be expressed as a simple polar equation,

$$r(\varphi) = a \cos\left(k\varphi + \gamma_0\right)$$

for any constant γ_0 (including 0). If k is an integer, these equations will produce a k-petaled rose if k is odd, or a $2k$-petaled rose if k is even. If k is rational but not an integer, a rose-like shape may form but with overlapping petals. Note that these equations never define a rose with 2, 6, 10, 14, etc. petals. The variable a directly represents the length or amplitude of the petals of the rose, while k relates to their spatial frequency. The constant γ_0 can be regarded as a phase angle.

Archimedean Spiral

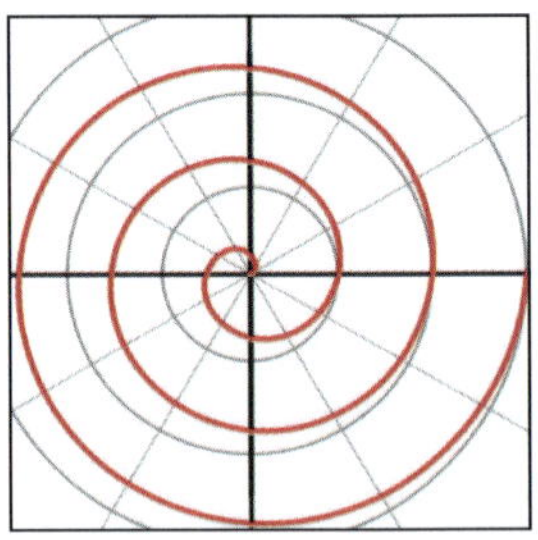

One arm of an Archimedean spiral
with equation $r(\varphi) = \varphi/2\pi$ for $0 < \varphi < 6\pi$.

The Archimedean spiral is a spiral that was discovered by Archimedes, which can also be expressed as a simple polar equation. It is represented by the equation,

$$r(\varphi) = a + b\varphi.$$

Changing the parameter a will turn the spiral, while b controls the distance between the arms, which for a given spiral is always constant. The Archimedean spiral has two arms, one for $\varphi > 0$ and one for $\varphi < 0$. The two arms are smoothly connected at the pole. Taking the mirror image of one arm across the 90°/270° line will yield the other arm. This curve is notable as one of the first curves, after the conic sections, to be described in a mathematical treatise, and as being a prime example of a curve that is best defined by a polar equation.

Conic Sections

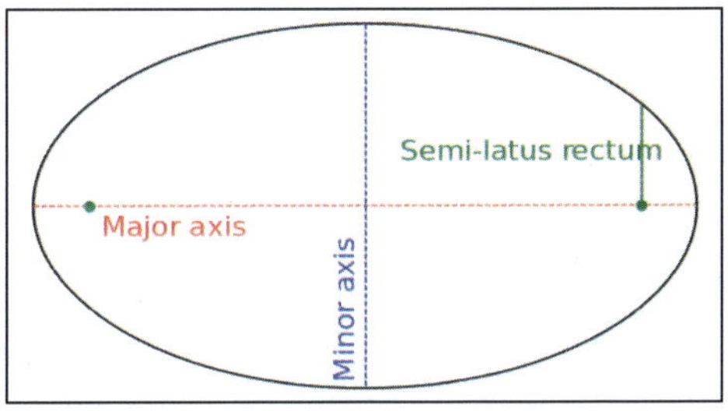

Ellipse, showing semi-latus rectum.

A conic section with one focus on the pole and the other somewhere on the 0° ray (so that the conic's major axis lies along the polar axis) is given by:

$$r = \frac{\ell}{1 - e\cos\varphi}$$

where e is the eccentricity and ℓ is the semi-latus rectum (the perpendicular distance at a focus from the major axis to the curve). If $e > 1$, this equation defines a hyperbola; if $e = 1$, it defines a parabola; and if $e < 1$, it defines an ellipse. The special case $e = 0$ of the latter results in a circle of the radius ℓ.

Intersection of Two Polar Curves

The graphs of two polar functions $r = f(\theta)$ and $r = g(\theta)$ have possible intersections of three types:

- In the origin if the equations $f(\theta) = 0$ and $g(\theta) = 0$ have at least one solution each.
- All the points $[g(\theta_i), \theta_i]$ where θ_i are the solutions to the equation $f(\theta + 2k\pi) = g(\theta)$ where k is an integer.
- All the points $[g(\theta_i), \theta_i]$ where θ_i are the solutions to the equation $f(\theta + (2k+1)\pi) = -g(\theta)$ where k is an integer.

Complex Numbers

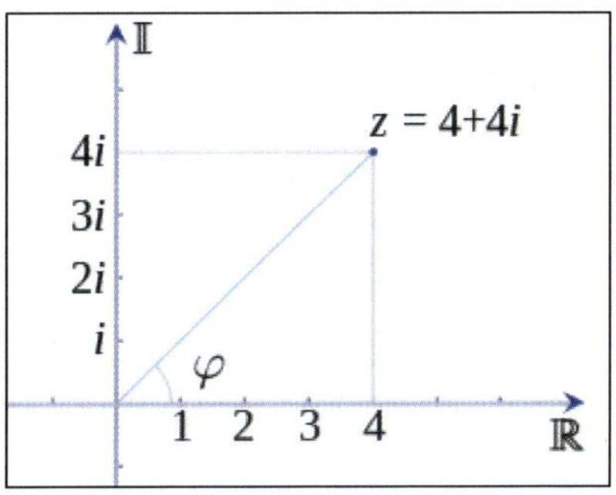

An illustration of a complex number z plotted on the complex plane.

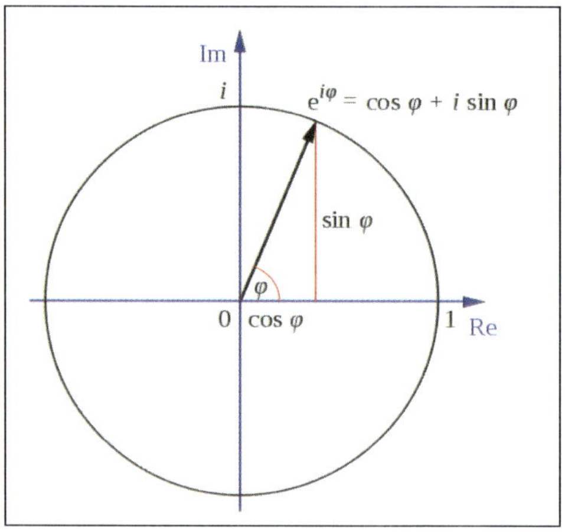

An illustration of a complex number plotted on the complex plane using Euler's formula.

Every complex number can be represented as a point in the complex plane, and can therefore be expressed by specifying either the point's Cartesian coordinates (called rectangular or Cartesian form) or the point's polar coordinates (called polar form). The complex number z can be represented in rectangular form as,

$$z = x + iy$$

where i is the imaginary unit, or can alternatively be written in polar form (via the conversion formulae given above) as,

$$z = r(\cos\varphi + i\sin\varphi)$$

and from there as,

$$z = re^{i\varphi}$$

where e is Euler's number, which are equivalent as shown by Euler's formula. (Note that this formula, like all those involving exponentials of angles, assumes that the angle φ is expressed in radians.) To convert between the rectangular and polar forms of a complex number, the conversion formulae given above can be used.

For the operations of multiplication, division, and exponentiation of complex numbers, it is generally much simpler to work with complex numbers expressed in polar form rather than rectangular form. From the laws of exponentiation:

Multiplication

$$r_0 e^{i\varphi_0} \, r_1 e^{i\varphi_1} = r_0 r_1 e^{i(\varphi_0 + \varphi_1)}$$

Division,

$$\frac{r_0 e^{i\varphi_0}}{r_1 e^{i\varphi_1}} = \frac{r_0}{r_1} e^{i(\varphi_0 - \varphi_1)}$$

Exponentiation (De Moivre's formula),

$$\left(re^{i\varphi}\right)^n = r^n e^{in\varphi}$$

Calculus

Calculus can be applied to equations expressed in polar coordinates.

The angular coordinate φ is expressed in radians throughout this section, which is the conventional choice when doing calculus.

Differential Calculus

Using $x = r\cos\varphi$ and $y = r\sin\varphi$, one can derive a relationship between derivatives in Cartesian and polar coordinates. For a given function, $u(x, y)$, it follows that (by computing its total derivatives):

$$r\frac{\partial u}{\partial r} = r\frac{\partial u}{\partial x}\frac{\partial x}{\partial r} + r\frac{\partial u}{\partial y}\frac{\partial y}{\partial r},$$

$$\frac{\partial u}{\partial \varphi} = \frac{\partial u}{\partial x}\frac{\partial x}{\partial \varphi} + \frac{\partial u}{\partial y}\frac{\partial y}{\partial \varphi},$$

Or

$$r\frac{\partial u}{\partial r} = r\frac{\partial u}{\partial x}\cos\varphi + r\frac{\partial u}{\partial y}\sin\varphi = x\frac{\partial u}{\partial x} + y\frac{\partial u}{\partial y},$$

$$\frac{\partial u}{\partial \varphi} = -\frac{\partial u}{\partial x}r\sin\varphi + \frac{\partial u}{\partial y}r\cos\varphi = -y\frac{\partial u}{\partial x} + x\frac{\partial u}{\partial y}.$$

Hence, we have the following formulae:

$$r\frac{\partial}{\partial r} = x\frac{\partial}{\partial x} + y\frac{\partial}{\partial y}$$

$$\frac{\partial}{\partial \varphi} = -y\frac{\partial}{\partial x} + x\frac{\partial}{\partial y}.$$

Using the inverse coordinates transformation, an analogous reciprocal relationship can be derived between the derivatives. Given a function $u(r, \varphi)$, it follows that,

$$\frac{\partial u}{\partial x} = \frac{\partial u}{\partial r}\frac{\partial r}{\partial x} + \frac{\partial u}{\partial \varphi}\frac{\partial \varphi}{\partial x},$$

$$\frac{\partial u}{\partial y} = \frac{\partial u}{\partial r}\frac{\partial r}{\partial y} + \frac{\partial u}{\partial \varphi}\frac{\partial \varphi}{\partial y},$$

Or

$$\frac{\partial u}{\partial x} = \frac{\partial u}{\partial r}\frac{x}{\sqrt{x^2 + y^2}} - \frac{\partial u}{\partial \varphi}\frac{y}{x^2 + y^2}$$

$$= \cos\varphi \frac{\partial u}{\partial r} - \frac{1}{r}\sin\varphi \frac{\partial u}{\partial \varphi},$$

$$\frac{\partial u}{\partial y} = \frac{\partial u}{\partial r}\frac{y}{\sqrt{x^2+y^2}} + \frac{\partial u}{\partial \varphi}\frac{x}{x^2+y^2}$$

$$= \sin\varphi \frac{\partial u}{\partial r} + \frac{1}{r}\cos\varphi \frac{\partial u}{\partial \varphi}.$$

Hence, we have the following formulae:

$$\frac{\partial}{\partial x} = \cos\varphi \frac{\partial}{\partial r} - \frac{1}{r}\sin\varphi \frac{\partial}{\partial \varphi}$$

$$\frac{\partial}{\partial y} = \sin\varphi \frac{\partial}{\partial r} + \frac{1}{r}\cos\varphi \frac{\partial}{\partial \varphi}.$$

To find the Cartesian slope of the tangent line to a polar curve $r(\varphi)$ at any given point, the curve is first expressed as a system of parametric equations.

$$x = r(\varphi)\cos\varphi$$

$$y = r(\varphi)\sin\varphi$$

Differentiating both equations with respect to φ yields,

$$\frac{dx}{d\varphi} = r'(\varphi)\cos\varphi - r(\varphi)\sin\varphi$$

$$\frac{dy}{d\varphi} = r'(\varphi)\sin\varphi + r(\varphi)\cos\varphi.$$

Dividing the second equation by the first yields the Cartesian slope of the tangent line to the curve at the point $(r(\varphi), \varphi)$:

$$\frac{dy}{dx} = \frac{r'(\varphi)\sin\varphi + r(\varphi)\cos\varphi}{r'(\varphi)\cos\varphi - r(\varphi)\sin\varphi}.$$

For other useful formulas including divergence, gradient, and Laplacian in polar coordinates.

Integral Calculus (Arc Length)

The arc length (length of a line segment) defined by a polar function is found by the integration over the curve $r(\varphi)$. Let L denote this length along the curve starting from points A through to point B, where these points correspond to $\varphi = a$ and $\varphi = b$ such that $0 < b - a < 2\pi$. The length of L is given by the following integral,

$$L = \int_a^b \sqrt{[r(\varphi)]^2 + \left[\tfrac{dr(\varphi)}{d\varphi}\right]^2}\, d\varphi$$

Integral Calculus (Area)

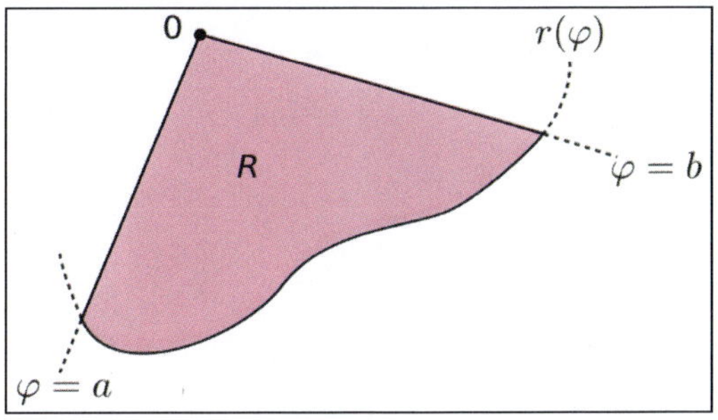

The integration region R is bounded by the curve $r(\varphi)$ and the rays $\varphi = a$ and $\varphi = b$.

Let R denote the region enclosed by a curve $r(\varphi)$ and the rays $\varphi = a$ and $\varphi = b$, where $0 < b - a \leq 2\pi$. Then, the area of R is,

$$\frac{1}{2}\int_a^b [r(\varphi)]^2 \, d\varphi.$$

This result can be found as follows. First, the interval $[a, b]$ is divided into n subintervals, where n is an arbitrary positive integer. Thus $\Delta\varphi$, the length of each subinterval, is equal to $b - a$ (the total length of the interval), divided by n, the number of subintervals. For each subinterval $i = 1, 2, ..., n$, let φi be the midpoint of the subinterval, and construct a sector with the center at the pole, radius $r(\varphi i)$, central angle $\Delta\varphi$ and arc length $r(\varphi i)\Delta\varphi$. The area of each constructed sector is therefore equal to,

$$[r(\varphi_i)]^2 \pi \cdot \frac{\Delta\varphi}{2\pi} = \frac{1}{2}[r(\varphi_i)]^2 \, \Delta\varphi.$$

Hence, the total area of all of the sectors is,

$$\sum_{i=1}^{n} \tfrac{1}{2} r(\varphi_i)^2 \, \Delta\varphi.$$

As the number of subintervals n is increased, the approximation of the area continues to improve. In the limit as $n \to \infty$, the sum becomes the Riemann sum for the above integral.

A mechanical device that computes area integrals is the planimeter, which measures the area of plane figures by tracing them out: this replicates integration in polar coordinates by adding a joint so that the 2-element linkage effects Green's theorem, converting the quadratic polar integral to a linear integral.

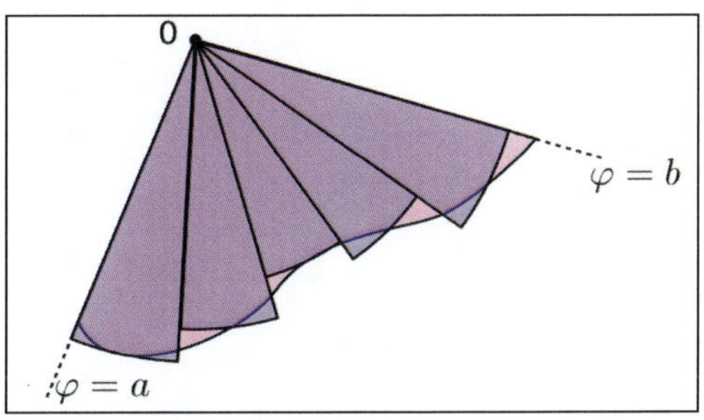

The region R is approximated by n sectors (here, $n = 5$).

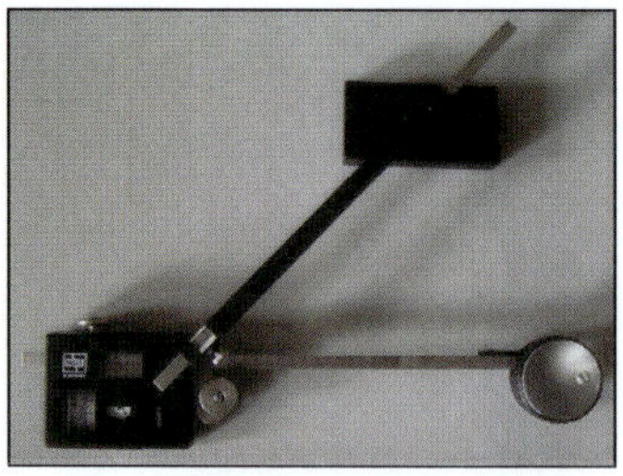

A planimeter, which mechanically computes polar integrals.

Generalization

Using Cartesian coordinates, an infinitesimal area element can be calculated as $dA = dx\,dy$. The substitution rule for multiple integrals states that, when using other coordinates, the Jacobian determinant of the coordinate conversion formula has to be considered:

$$J = \det\frac{\partial(x,y)}{\partial(r,\varphi)} = \begin{vmatrix} \frac{\partial x}{\partial r} & \frac{\partial x}{\partial \varphi} \\ \frac{\partial y}{\partial r} & \frac{\partial y}{\partial \varphi} \end{vmatrix} = \begin{vmatrix} \cos\varphi & -r\sin\varphi \\ \sin\varphi & r\cos\varphi \end{vmatrix} = r\cos^2\varphi + r\sin^2\varphi = r.$$

Hence, an area element in polar coordinates can be written as,

$$dA = dx\,dy = J\,dr\,d\varphi = r\,dr\,d\varphi.$$

Now, a function, that is given in polar coordinates, can be integrated as follows:

$$\iint_R f(x,y)\,dA = \int_a^b \int_0^{r(\varphi)} f(r,\varphi)\,r\,dr\,d\varphi.$$

Here, R is the same region as above, namely, the region enclosed by a curve $r(\phi)$ and the rays $\varphi = a$ and $\varphi = b$. The formula for the area of R mentioned above is retrieved by taking f identically equal to 1.

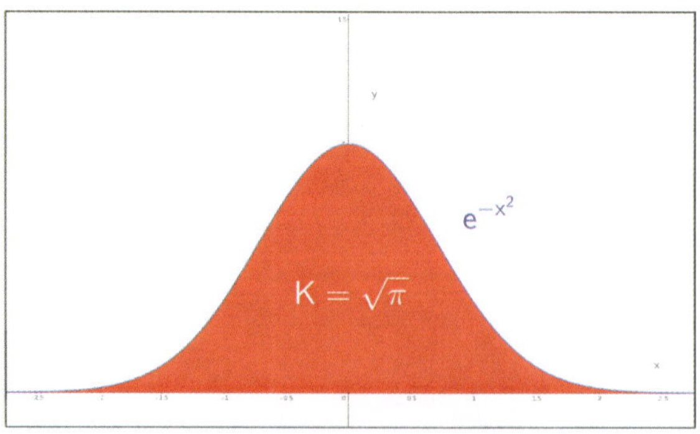

The Gaussian integral.

A more surprising application of this result yields the Gaussian integral, here denoted K:

$$K = \int_{-\infty}^{\infty} e^{-x^2}\,dx = \sqrt{\pi}.$$

Vector Calculus

Vector calculus can also be applied to polar coordinates. For a planar motion, let $\mathbf{r}$ be the position vector ($r\cos(\varphi)$, $r\sin(\varphi)$), with r and φ depending on time t.

We define the unit vectors

$$\hat{\mathbf{r}} = (\cos(\varphi), \sin(\varphi))$$

in the direction of $\mathbf{r}$ and,

$$\hat{\boldsymbol{\varphi}} = (-\sin(\varphi), \cos(\varphi)) = \hat{\mathbf{k}} \times \hat{\mathbf{r}},$$

in the plane of the motion perpendicular to the radial direction, where $\hat{\mathbf{k}}$ is a unit vector normal to the plane of the motion.

Then,

$$\mathbf{r} = (x, y) = r(\cos\varphi, \sin\varphi) = r\hat{\mathbf{r}},$$

$$\dot{\mathbf{r}} = (\dot{x}, \dot{y}) = \dot{r}(\cos\varphi, \sin\varphi) + r\dot{\varphi}(-\sin\varphi, \cos\varphi) = \dot{r}\hat{\mathbf{r}} + r\dot{\varphi}\hat{\varphi},$$

$$\begin{aligned}\ddot{\mathbf{r}} &= (\ddot{x}, \ddot{y}) \\ &= \ddot{r}(\cos\varphi, \sin\varphi) + 2\dot{r}\dot{\varphi}(-\sin\varphi, \cos\varphi) + r\ddot{\varphi}(-\sin\varphi, \cos\varphi) - r\dot{\varphi}^2(\cos\varphi, \sin\varphi) \\ &= \left(\ddot{r} - r\dot{\varphi}^2\right)\hat{\mathbf{r}} + \left(r\ddot{\varphi} + 2\dot{r}\dot{\varphi}\right)\hat{\boldsymbol{\varphi}} \\ &= \left(\ddot{r} - r\dot{\varphi}^2\right)\hat{\mathbf{r}} + \frac{1}{r}\frac{d}{dt}\left(r^2\dot{\varphi}\right)\hat{\boldsymbol{\varphi}}.\end{aligned}$$

Centrifugal and Coriolis Terms

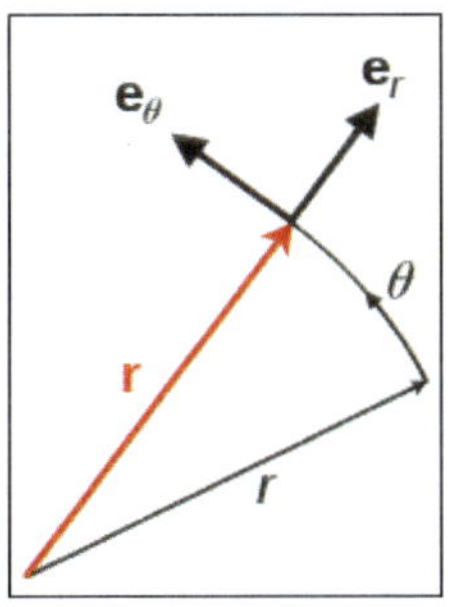

Position vector **r**, always points radially from the origin.

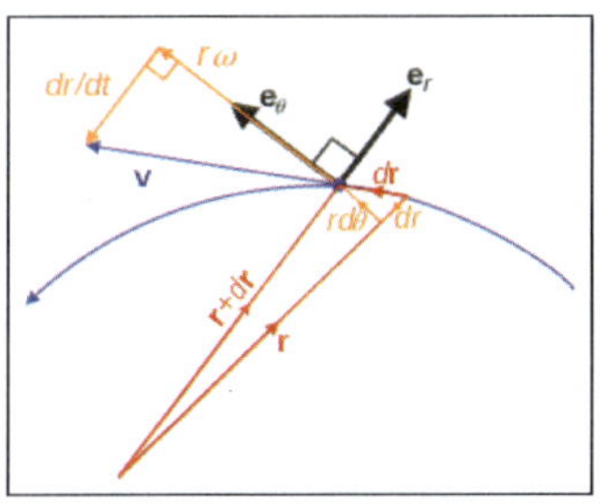

Velocity vector **v**, always tangent to the path of motion.

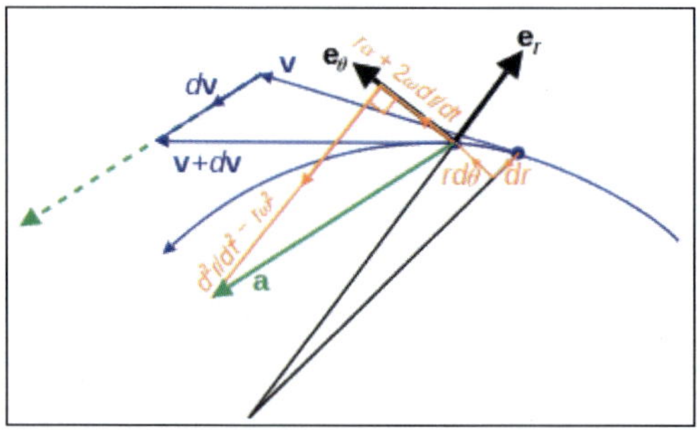

Acceleration vector **a**, not parallel to the radial motion but offset by the angular and Coriolis accelerations, nor tangent to the path but offset by the centripetal and radial accelerations.

Kinematic vectors in plane polar coordinates. Notice the setup is not restricted to 2d space, but a plane in any higher dimension.

The term $r\dot{\varphi}^2$ is sometimes referred to as the *centripetal acceleration*, and the term $2\dot{r}\dot{\varphi}$ as the *Coriolis acceleration*.

Note: these terms, that appear when acceleration is expressed in polar coordinates, are a mathematical consequence of differentiation; they appear whenever polar coordinates are used. In planar particle dynamics these accelerations appear when setting up Newton's second law of motion in a rotating frame of reference. Here these extra terms are often called fictitious forces; fictitious because they are simply a result of a change in coordinate frame. That does not mean they do not exist, rather they exist only in the rotating frame.

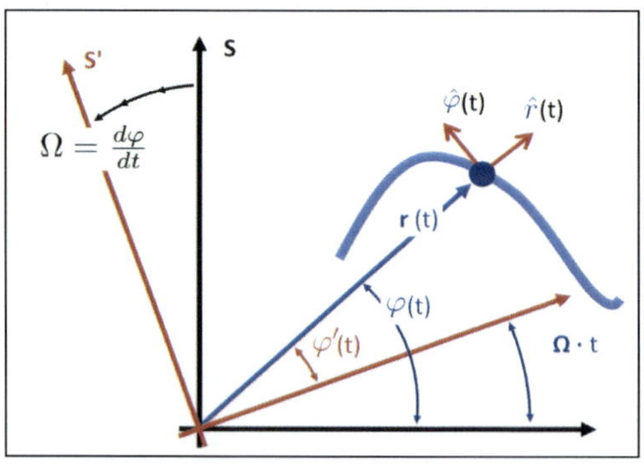

Inertial frame of reference S and instantaneous non-inertial co-rotating frame of reference S'. The co-rotating frame rotates at angular rate Ω equal to the rate of rotation of the particle about the origin of S' at the particular moment t. Particle is located at vector position $r(t)$ and unit vectors are shown in the radial direction to the particle from the origin, and also in the direction of increasing angle ϕ normal to the radial direction. These unit vectors need not be related to the tangent and normal to the path. Also, the radial distance r need not be related to the radius of curvature of the path.

Co-rotating Frame

For a particle in planar motion, one approach to attaching physical significance to these terms is based on the concept of an instantaneous *co-rotating frame of reference*. To define a co-rotating frame, first an origin is selected from which the distance $r(t)$ to the particle is defined. An axis of rotation is set up that is perpendicular to the plane of motion of the particle, and passing through this origin. Then, at the selected moment *t*, the rate of rotation of the co-rotating frame Ω is made to match the rate of rotation of the particle about this axis, $d\varphi/dt$. Next, the terms in the acceleration in the inertial frame are related to those in the co-rotating frame. Let the location of the particle in the inertial frame be (*r*(t), φ(t)), *and in the co-rotating frame be* (r(t), φ′(t)). *Because the co-rotating frame rotates at the same rate as the particle,* $d\varphi'/dt = 0$. *The fictitious centrifugal force in the co-rotating frame is* $mr\Omega^2$, radially outward. The velocity of the particle in the co-rotating frame also is radially outward, because $d\varphi'/dt = 0$. The *fictitious Coriolis force* therefore has a value $-2m(dr/dt)\Omega$, pointed in the direction of increasing φ only. Thus, using these forces in Newton's second law we find:

$$\mathbf{F} + \mathbf{F}_{\text{cf}} + \mathbf{F}_{\text{Cor}} = m\ddot{\mathbf{r}} ,$$

where over dots represent time differentiations, and **F** is the net real force (as opposed to the fictitious forces). In terms of components, this vector equation becomes:

$$F_r + mr\Omega^2 = m\ddot{r}$$
$$F_\varphi - 2m\dot{r}\Omega = mr\ddot{\varphi} ,$$

which can be compared to the equations for the inertial frame:

$$F_r = m\ddot{r} - mr\dot{\varphi}^2$$
$$F_\varphi = mr\ddot{\varphi} + 2m\dot{r}\dot{\varphi} .$$

This comparison, plus the recognition that by the definition of the co-rotating frame at time *t* it has a rate of rotation $\Omega = d\varphi/dt$, shows that we can interpret the terms in the acceleration (multiplied by the mass of the particle) as found in the inertial frame as the negative of the centrifugal and Coriolis forces that would be seen in the instantaneous, non-inertial co-rotating frame.

For general motion of a particle (as opposed to simple circular motion), the centrifugal and Coriolis forces in a particle's frame of reference commonly are referred to the instantaneous osculating circle of its motion, not to a fixed center of polar coordinates.

Differential Geometry

In the modern terminology of differential geometry, polar coordinates provide coordinate charts for the differentiable manifold $\mathbb{R}^2 \setminus \{(0,0)\}$, the plane minus the origin. In these coordinates, the Euclidean metric tensor is given by,

$$ds^2 = dr^2 + r^2 d\theta^2 .$$

This can be seen via the change of variables formula for the metric tensor, or by computing the differential forms *dx*, *dy* via the exterior derivative of the 0-forms $x = r\cos(\theta)$, $y = r\sin(\theta)$ and substituting them in the Euclidean metric tensor $ds^2 = dx^2 + dy^2$. An orthonormal frame with respect to this metric is given by,

$$e_r = \frac{\partial}{\partial r}, \quad e_\theta = \frac{1}{r}\frac{\partial}{\partial \theta},$$

with dual coframe,

$$e^r = dr, \quad e^\theta = r d\theta.$$

The connection form relative to this frame and the Levi-Civita connection is given by the skew-symmetric matrix of 1-forms,

$$\omega^i_j = \begin{pmatrix} 0 & -d\theta \\ d\theta & 0 \end{pmatrix}$$

and hence the curvature form $\Omega = d\omega + \omega \wedge \omega$ vanishes identically. Therefore, as expected, the punctured plane is a flat manifold.

Extensions in 3D

The polar coordinate system is extended into three dimensions with two different coordinate systems, the cylindrical and spherical coordinate system.

Applications

Polar coordinates are two-dimensional and thus they can be used only where point positions lie on a single two-dimensional plane. They are most appropriate in any context where the phenomenon being considered is inherently tied to direction and length from a center point. For instance, the examples above show how elementary polar equations suffice to define curves—such as the Archimedean spiral—whose equation in the Cartesian coordinate system would be much more intricate. Moreover, many physical systems—such as those concerned with bodies moving around a central point or with phenomena originating from a central point—are simpler and more intuitive to model using polar coordinates. The initial motivation for the introduction of the polar system was the study of circular and orbital motion.

Position and Navigation

Polar coordinates are used often in navigation as the destination or direction of travel can be given as an angle and distance from the object being considered. For instance, aircraft use a slightly modified version of the polar coordinates for navigation. In this system, the one generally used for any sort of navigation, the 0° ray is generally called heading 360, and the angles continue in a clockwise direction, rather than counterclockwise, as in the mathematical system. Heading 360 corresponds to magnetic north, while headings 90, 180, and 270 correspond to magnetic east, south, and west, respectively. Thus, an aircraft traveling 5 nautical miles due east will be traveling 5 units at heading 90 (read zero-niner-zero by air traffic control).

Modeling

Systems displaying radial symmetry provide natural settings for the polar coordinate system, with the central point acting as the pole. A prime example of this usage is the groundwater flow equation when applied to radially symmetric wells. Systems with a radial force are also good candidates for the use of the polar coordinate system. These systems include gravitational fields, which obey the inverse-square law, as well as systems with point sources, such as radio antennas.

Radially asymmetric systems may also be modeled with polar coordinates. For example, a microphone's pickup pattern illustrates its proportional response to an incoming sound from a given direction, and these patterns can be represented as polar curves. The curve for a standard cardioid microphone, the most common unidirectional microphone, can be represented as $r = 0.5 + 0.5\sin(\phi)$ at its target design frequency. The pattern shifts toward omnidirectionality at lower frequencies.

Cylindrical Coordinate System

A cylindrical coordinate system is a three-dimensional coordinate system that specifies point positions by the distance from a chosen reference axis, the direction from the axis relative to a chosen reference direction, and the distance from a chosen reference plane perpendicular to the axis. The latter distance is given as a positive or negative number depending on which side of the reference plane faces the point.

The *origin* of the system is the point where all three coordinates can be given as zero. This is the intersection between the reference plane and the axis. The axis is variously called the *cylindrical* or *longitudinal* axis, to differentiate it from the *polar axis*, which is the ray that lies in the reference plane, starting at the origin and pointing in the reference direction. Other directions perpendicular to the longitudinal axis are called *radial lines*.

The distance from the axis may be called the *radial distance* or *radius*, while the angular coordinate is sometimes referred to as the *angular position* or as the *azimuth*.The radius and the azimuth are together called the *polar coordinates*, as they correspond to a two-dimensional polar coordinate system in the plane through the point, parallel to the reference plane. The third coordinate may be called the *height* or *altitude* (if the reference plane is considered horizontal), *longitudinal position*, or *axial position*.

Cylindrical coordinates are useful in connection with objects and phenomena that have some rotational symmetry about the longitudinal axis, such as water flow in a straight pipe with round cross-section, heat distribution in a metal cylinder, electromagnetic fields produced by an electric current in a long, straight wire, accretion disks in astronomy, and so on.

They are sometimes called "cylindrical polar coordinates" and "polar cylindrical coordinates", and are sometimes used to specify the position of stars in a galaxy ("galactocentric cylindrical polar coordinates").

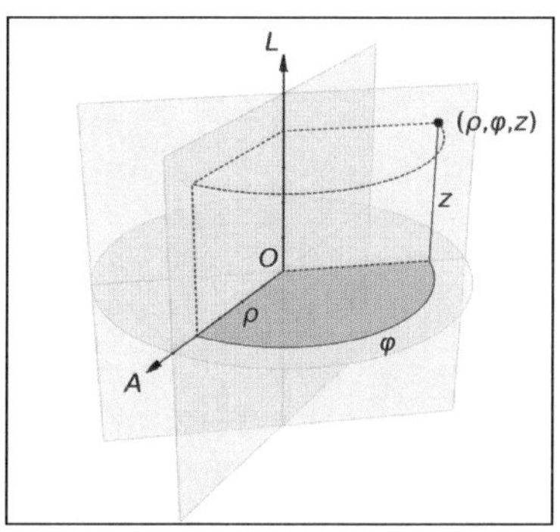

A cylindrical coordinate system with origin O, polar axis A, and longitudinal axis L. The dot is the point with radial distance $\rho = 4$, angular coordinate $\varphi = 130°$, and height $z = 4$.

The three coordinates (ρ, φ, z) of a point P are defined as:

- The *axial distance* or *radial distance* ρ is the Euclidean distance from the z-axis to the point P.
- The *azimuth* φ is the angle between the reference direction on the chosen plane and the line from the origin to the projection of P on the plane.
- The *axial coordinate* or *height* z is the signed distance from the chosen plane to the point P.

Unique Cylindrical Coordinates

As in polar coordinates, the same point with cylindrical coordinates (ρ, φ, z) has infinitely many equivalent coordinates, namely (ρ, $\varphi \pm n\times360°$, z) and ($-\rho$, $\varphi \pm (2n + 1)\times180°$, z), where n is any integer. Moreover, if the radius ρ is zero, the azimuth is arbitrary.

In situations where someone wants a unique set of coordinates for each point, one may restrict the radius to be non-negative ($\rho \geq 0$) and the azimuth φ to lie in a specific interval spanning 360°, such as [−180°,+180°] or [0,360°].

Conventions

The notation for cylindrical coordinates is not uniform. The ISO standard 31-11 recommends (ρ, φ, z), where ρ is the radial coordinate, φ the azimuth, and z the height. However, the radius is also often denoted r or s, the azimuth by θ or t, and the third coordinate by h or (if the cylindrical axis is considered horizontal) x, or any context-specific letter.

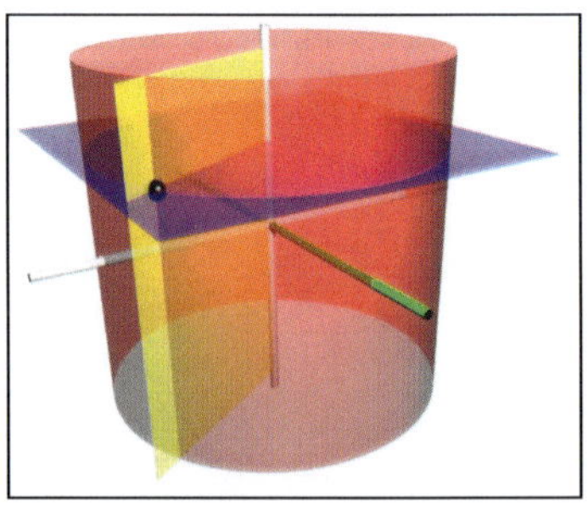

The coordinate surfaces of the cylindrical coordinates (ρ, φ, z). The red cylinder shows the points with $\rho = 2$, the blue plane shows the points with $z = 1$, and the yellow half-plane shows the points with $\varphi = -60°$. The z-axis is vertical and the x-axis is highlighted in green. The three surfaces intersect at the point P with those coordinates (shown as a black sphere); the Cartesian coordinates of P are roughly (1.0, −1.732, 1.0).

In concrete situations, and in many mathematical illustrations, a positive angular coordinate is measured counterclockwise as seen from any point with positive height.

Coordinate System Conversions

The cylindrical coordinate system is one of many three-dimensional coordinate systems. The following formulae may be used to convert between them.

Cartesian Coordinates

For the conversion between cylindrical and Cartesian coordinates, it is convenient to assume that the reference plane of the former is the Cartesian *xy*-plane (with equation $z = 0$), and the cylindrical axis is the Cartesian *z*-axis. Then the *z*-coordinate is the same in both systems, and the correspondence between cylindrical (ρ,φ,z) and Cartesian (x,y,z) are the same as for polar coordinates, namely,

$$x = \rho \cos \varphi$$
$$y = \rho \sin \varphi$$
$$z = z$$

in one direction, and

$$\rho = \sqrt{x^2 + y^2}$$

$$\varphi = \begin{cases} 0 & \text{if } x = 0 \text{ and } y = 0 \\ \arcsin\left(\dfrac{y}{\rho}\right) & \text{if } x \geq 0 \\ \arctan\left(\dfrac{y}{x}\right) & \text{if } x > 0 \\ -\arcsin\left(\dfrac{y}{\rho}\right) + \pi & \text{if } x < 0 \end{cases}$$

in the other. The arcsin function is the inverse of the sine function, and is assumed to return an angle in the range $\left[-\frac{\pi}{2}, +\frac{\pi}{2}\right] = [-90°, +90°]$. These formulas yield an azimuth φ in the range $[-90°, +270°]$.

Many modern programming languages provide a function that will compute the correct azimuth φ, in the range $(-\pi, \pi)$, given x and y, without the need to perform a case analysis as above. For example, this function is called by atan2(*y*, *x*) in the C programming language, and atan(*y*, *x*) in Common Lisp.

Spherical Coordinates

Spherical coordinates (radius r, elevation or inclination θ, azimuth φ), may be converted into cylindrical coordinates by:

θ *is elevation:*	θ *is inclination:*
$\rho = r \cos \theta$	$\rho = r \sin \theta$
$\varphi = \varphi$	$\varphi = \varphi$
$z = r \sin \theta$	$z = r \cos \theta$

Cylindrical coordinates may be converted into spherical coordinates by:

θ *is elevation:*	θ *is inclination:*
$r = \sqrt{\rho^2 + z^2}$ $\theta = \arctan\left(\frac{z}{\rho}\right)$ $\varphi = \varphi$	$r = \sqrt{\rho^2 + z^2}$ $\theta = \arctan\left(\frac{\rho}{z}\right)$ $\varphi = \varphi$

Line and Volume Elements

In many problems involving cylindrical polar coordinates, it is useful to know the line and volume elements; these are used in integration to solve problems involving paths and volumes.

The line element is,

$$\mathrm{d}\mathbf{r} = \mathrm{d}\rho\,\hat{\boldsymbol{\rho}} + \rho\,\mathrm{d}\varphi\,\hat{\boldsymbol{\varphi}} + \mathrm{d}z\,\hat{\mathbf{z}}.$$

The volume element is

$$\mathrm{d}V = \rho\,\mathrm{d}\rho\,\mathrm{d}\varphi\,\mathrm{d}z.$$

The surface element in a surface of constant radius ρ (a vertical cylinder) is,

$$\mathrm{d}S_\rho = \rho\,\mathrm{d}\varphi\,\mathrm{d}z.$$

The surface element in a surface of constant azimuth φ (a vertical half-plane) is,

$$\mathrm{d}S_\varphi = \mathrm{d}\rho\,\mathrm{d}z.$$

The surface element in a surface of constant height z (a horizontal plane) is,

$$\mathrm{d}S_z = \rho\,\mathrm{d}\rho\,\mathrm{d}\varphi.$$

The del operator in this system leads to the following expressions for gradient, divergence, curl and Laplacian:

$$\nabla f = \frac{\partial f}{\partial \rho}\hat{\boldsymbol{\rho}} + \frac{1}{\rho}\frac{\partial f}{\partial \varphi}\hat{\boldsymbol{\varphi}} + \frac{\partial f}{\partial z}\hat{\mathbf{z}}$$

$$\nabla \cdot \mathbf{A} = \frac{1}{\rho}\frac{\partial}{\partial \rho}\left(\rho A_\rho\right) + \frac{1}{\rho}\frac{\partial A_\varphi}{\partial \varphi} + \frac{\partial A_z}{\partial z}$$

$$\nabla \times \mathbf{A} = \left(\frac{1}{\rho}\frac{\partial A_z}{\partial \varphi} - \frac{\partial A_\varphi}{\partial z}\right)\hat{\boldsymbol{\rho}} + \left(\frac{\partial A_\rho}{\partial z} - \frac{\partial A_z}{\partial \rho}\right)\hat{\boldsymbol{\varphi}} + \frac{1}{\rho}\left(\frac{\partial}{\partial \rho}\left(\rho A_\varphi\right) - \frac{\partial A_\rho}{\partial \varphi}\right)\hat{\mathbf{z}}$$

$$\nabla^2 f = \frac{1}{\rho}\frac{\partial}{\partial \rho}\left(\rho\frac{\partial f}{\partial \rho}\right) + \frac{1}{\rho^2}\frac{\partial^2 f}{\partial \varphi^2} + \frac{\partial^2 f}{\partial z^2}$$

Cylindrical Harmonics

The solutions to the Laplace equation in a system with cylindrical symmetry are called cylindrical harmonics.

Spherical Coordinate System

Spherical coordinates (r, θ, φ) as commonly used in physics: radial distance r, polar angle θ (theta), and azimuthal angle φ (phi). The symbol ρ (rho) is often used instead of r.

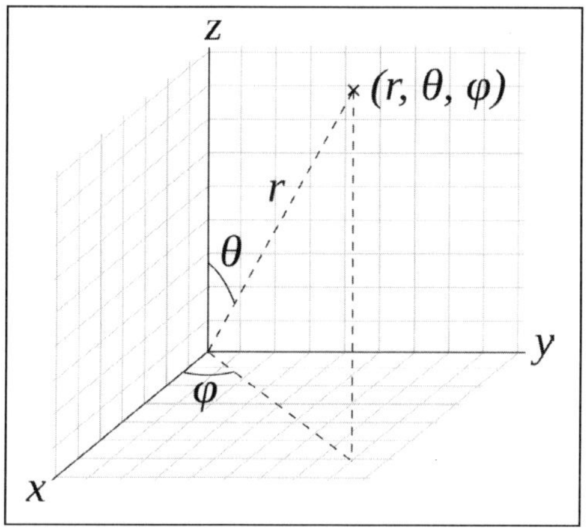

Coordinate System Conversions

As the spherical coordinate system is only one of many three-dimensional coordinate systems, there exist equations for converting coordinates between the spherical coordinate system and others.

Cartesian Coordinates

The spherical coordinates of a point in the ISO convention (radius r, inclination θ, azimuth φ) can be obtained from its Cartesian coordinates (x, y, z) by the formulae,

$$r = \sqrt{x^2 + y^2 + z^2}$$

$$\theta = \arccos\left(\frac{z}{\sqrt{x^2 + y^2 + z^2}}\right) = \arccos\left(\frac{z}{r}\right)$$

$$\varphi = \arctan\left(\frac{x}{y}\right)$$

Conversely, the Cartesian coordinates may be retrieved from the spherical coordinates (radius r, inclination θ, azimuth φ), where r $\in [0, \infty)$, $\theta \in [0, \pi]$, $\varphi \in [0, 2\pi)$, by:

$$x = r \sin\theta \cos\varphi$$

$$y = r \sin\theta \sin\varphi$$

$$z = r \cos\theta$$

Spherical Coordinate System as Commonly used in Mathematics

Spherical coordinates (r, θ, φ) as often used in mathematics: radial distance r, azimuthal angle θ, and polar angle φ. The meanings of θ and φ have been swapped compared to the physics convention.

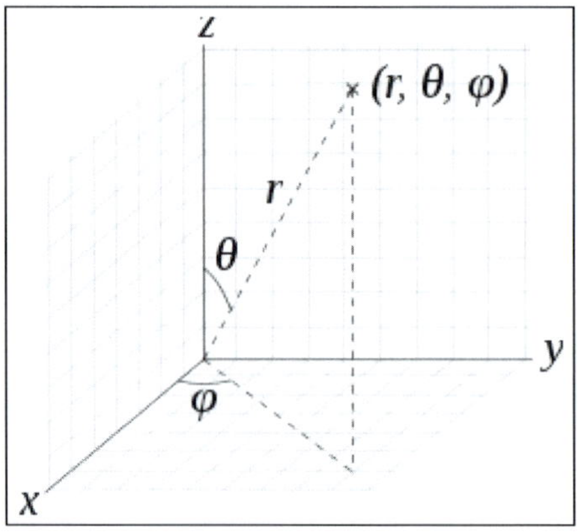

Conic Sections

In mathematics, a conic section (or just conic) is a curve that can be formed by intersecting a cone (more precisely, a right circular conical surface) with a plane. The conic sections were named and studied as long ago as 200 B.C.E., when Apollonius of Perga undertook a systematic study of their properties.

Two well-known conics are the circle and the ellipse. They arise when the intersection of the cone and plane is a closed curve. The circle is a special case of the ellipse in which the plane is perpendicular to the axis of the cone. If the plane is parallel to a generator line of the cone, the conic is called a parabola. Finally, if the intersection is an open curve and the plane is not parallel to generator lines of the cone, the figure is a hyperbola. (In this case the plane will intersect both halves of the cone, producing two separate curves, though often one is ignored.)

Conic sections are observed in the paths taken by celestial bodies. When two massive objects interact according to Newton's law of universal gravitation, their orbits are conic sections if their common center of mass is considered to be at rest. If they are bound together, they will both trace out ellipses; if they are moving apart, they will both follow parabolas or hyperbolas.

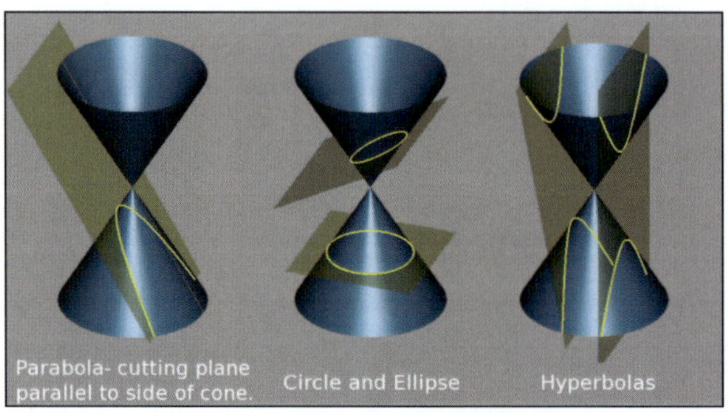

Types of conic sections.

The study of conic sections is important not only for mathematics, physics, and astronomy, but also for a variety of engineering applications. The smoothness of conic sections is an important property for applications such as aerodynamics, where a smooth surface is needed to ensure laminar flow and prevent turbulence.

Degenerate Cases

There are a number of degenerate cases, in which the plane passes through the apex of the cone. The intersection in these cases can be a straight line (when the plane is tangential to the surface of the cone); a point (when the angle between the plane and the axis of the cone is larger than this); or a pair of intersecting lines (when the angle is smaller). There is also a degenerate where the cone is a cylinder (the vertex is at infinity), which can produce two parallel lines.

Eccentricity

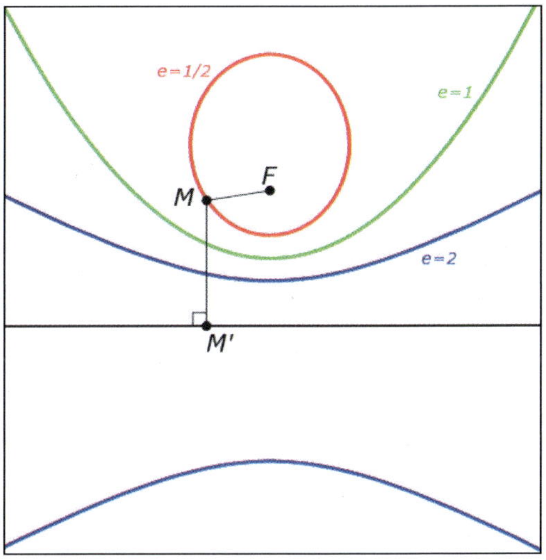

Ellipse ($e = 1/2$), parabola ($e = 1$) and hyperbola ($e = 2$) with fixed focus F and directrix.

The four defining conditions above can be combined into one condition that depends on a fixed point F (the focus), a line L (the directrix) not containing F, and a nonnegative real number e (the eccentricity). The corresponding conic section consists of all points whose distance to F equals e times their distance to L. For 0 < e < 1 an ellipse is obtained; for e = 1, a parabola; and for e > 1, a hyperbola.

For an ellipse and a hyperbola, two focus-directrix combinations can be taken, each giving the same full ellipse or hyperbola. The distance from the center to the directrix is a/e, where a is the semi-major axis of the ellipse, or the distance from the center to the tops of the hyperbola. The distance from the center to a focus is ae.

In the case of a circle, the eccentricity e = 0, and one can imagine the directrix to be infinitely far removed from the center. However, the statement that the circle consists of all points whose distance is e times the distance to L is not useful, because it yields zero times infinity.

The eccentricity of a conic section is thus a measure of how far it deviates from being circular. For a given a, the closer e is to 1, the smaller is the semi-minor axis.

Hyperbola

Hyperbolas are one of the four conic sections, and are described by certain kinds of equations. A hyperbola can be defined in a number of ways. A hyperbola is:

- The intersection of a right circular double cone with a plane at an angle greater than the slope of the cone (for example, perpendicular to the base of the cone).
- The set of all points such that the difference between the distances to two focal points is constant.
- The set of all points such that the ratio of the distance to a single focal point divided by the distance to a line (the directrix) is greater than one.

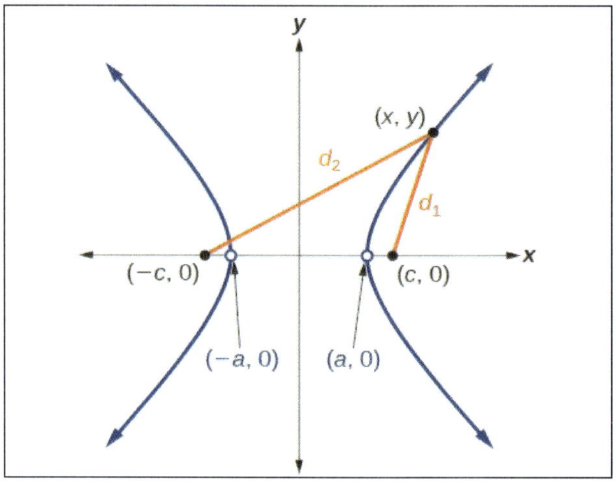

Diagram of a hyperbola: The hyperbola, shown in blue, has a center at the origin, two focal points at (−c, 0) and (c, 0), and two vertices located at + a and – a on the x-axis.

We begin with two focal points, F_1 and F_2, located on the xx-axis, so that they have coordinates (c, 0) and (−c, 0) (other arrangements are possible). We want the set of all points that have the same *difference* between the distances to these points. The center of this hyperbola is the origin (0, 0).

Imagine that we take a point on the red hyperbola curve, called P, and we let that point be the + a value on the x-axis. Then the difference of distances between P and the two focal points is:

$$\begin{aligned}(P \rightarrow F_2) - (P \rightarrow F_1) &= (c+a) - (c-a) \\ &= 2a\end{aligned}$$

where a is the distance from the center (origin) to the vertices of the hyperbola. With this value for the difference of distances, we can choose any point (x, y) on the hyperbola and construct an equation by use of the distance formula:

$$\sqrt{(x-c)^2+(y-0)^2} - \sqrt{(x-(-c))^2+(y-0)^2} = 2a$$

$$\sqrt{(x-c)^2+y^2} - \sqrt{(x+c)^2+y^2} = 2a$$

From here there is some straightforward, but messy, algebra. We need to square both sides of this equation multiple times if we want the variables to escape their square roots. When the dust settles, we have:

$$x^2(c^2-a^2)-a^2y^2=a^2(c^2-a^2)$$

At this point we introduce one more parameter, defined as $b^2=c^2-a^2$, which reduces the hyperbola even further:

$$b^2x^2-a^2y^2=a^2b^2$$

Lastly we divide both sides of the equation by a^2b^2:

$$b^2x^2-a^2y^2=a^2b^2$$

$$\frac{b^2x^2}{a^2b^2}-\frac{a^2y^2}{a^2b^2}=\frac{a^2b^2}{a^2b^2}$$

$$\frac{x^2}{a^2}-\frac{y^2}{b^2}=1$$

Thus, the standard form of the equation for a hyperbola with focal points on the x axis is:

$$\frac{x^2}{a^2}-\frac{y^2}{b^2}=1$$

If the focal points are on the y-axis, the variables simply change places:

$$\frac{y^2}{a^2}-\frac{x^2}{b^2}=1$$

Note that the hyperbola standard form is very similar to the standard form of the ellipse:

$$\frac{x^2}{a^2}+\frac{y^2}{b^2}=1$$

The similarity is not coincidental. The ellipse can be defined as all points that have a constant sum of distances to two focal points, and the hyperbola is defined as all points that have constant difference of distances to two focal points.

There is another common form of hyperbola equation that, at first glance, looks very different:

$$y=\frac{1}{x} \text{ or } xy = 1.$$

From the graph, it can be seen that the hyperbola formed by the equation xy = 1 is the same shape as the standard form hyperbola, but rotated by 45°. To prove that it is the same as the standard hyperbola, you can check for yourself that it has two focal points and that all points have the same difference of distances. Another way to prove it algebraically is to construct a rotated xx–y coordinate frame.

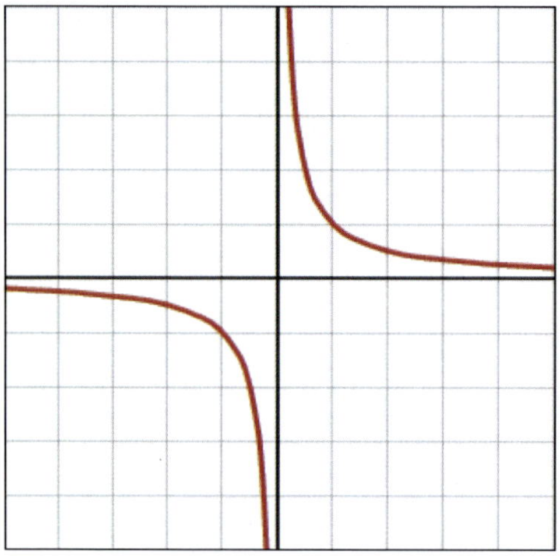

Reciprocal hyperbola: This hyperbola is defined by the equation xy = 1.

Parts of a Hyperbola

The features of a hyperbola can be determined from its equation.

A hyperbola is one of the four conic sections. All hyperbolas share common features, and it is possible to determine the specifics of any hyperbola from the equation that defines it.

Standard Form

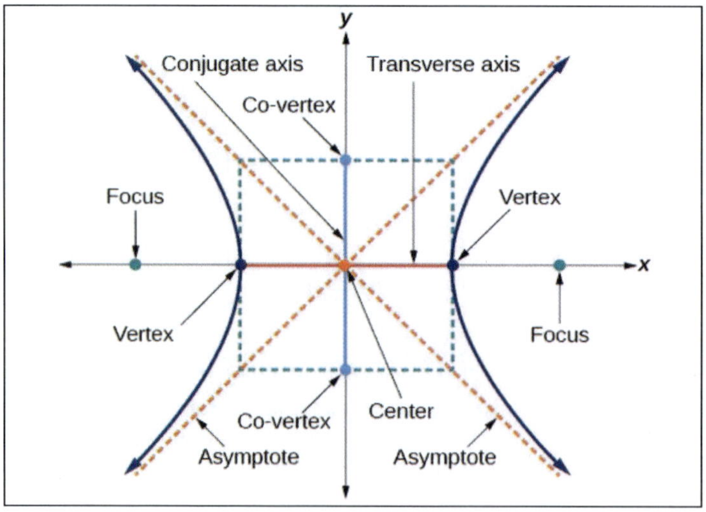

Diagram of a hyperbola: All hyperbolas share common features. A hyperbola consists of two curves, each with a vertex and a focus. The transverse axis is the axis that crosses through both vertices and foci, and the conjugate axis is perpendicular to it. A hyperbola also has asymptotes which cross in an "x". The two branches of the hyperbola are on opposite sides of the asymptotes' cross. The vertices and asymptotes can be used to form a rectangle, with the vertices at the centers of two opposite sides and the corners on the asymptotes. The centers of the other two sides, along the conjugate axis, are called the co-vertices. Where the asymptotes of the hyperbola cross is called the center.

If the foci lie on the x-axis, the standard form of a hyperbola is:

$$\frac{(x-h)^2}{a^2}-\frac{(y-k)^2}{b^2}=1$$

If the foci lie on the y-axis, the standard form is:

$$\frac{(y-k)^2}{a^2}-\frac{(x-h)^2}{b^2}=1$$

We will use the x-axis hyperbola to demonstrate how to determine the features of a hyperbola, so that a is associated with x-coordinates and b is associated with y-coordinates. For a y-axis hyperbola, the associations are reversed.

Center

The center has coordinates (h, k).

Vertices

The vertices have coordinates (h + a, k) and (h −a, k). The line connecting the vertices is called the transverse axis.

Co-Vertices

The co-vertices correspond to b, the "minor semi-axis length", and have coordinates (h, k + b) and (h, k – b).

Asymptotes

The major and minor axes a and b, as the vertices and co-vertices, describe a rectangle that shares the same center as the hyperbola, and has dimensions 2a × 2b. The asymptotes of the hyperbola are straight lines that are the diagonals of this rectangle. We can therefore use the corners of the rectangle to define the equation of these lines:

$$y=\pm\frac{a}{b}(x-h)+k$$

The rectangle itself is also useful for drawing the hyperbola graph by hand, as it contains the vertices. When drawing the hyperbola, draw the rectangle first. Then draw in the asymptotes as extended lines that are also the diagonals of the rectangle. Finally, draw the curve of the hyperbola by following the asymptote inwards, curving in to touch the vertex on the rectangle, and then following the other asymptote out. Repeat for the other branch.

Focal Points

The foci have coordinates (h + c, k) and (h – c, k). The value of cc is found with this relation:

$$c^2+a^2+b^2$$

Rectangular Hyperbola

Rectangular hyperbolas, defined by,

$$(x-h)(y-k)=m$$

for some constant m, are much simpler to analyze than standard form hyperbolas.

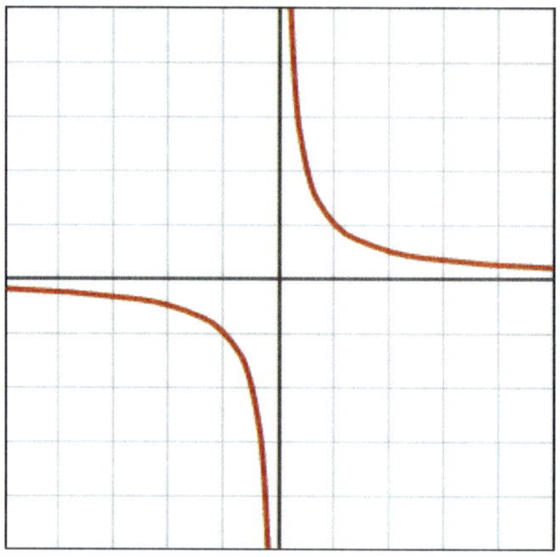

Rectangular hyperbola: This rectangular hyperbola has its center at the origin, and is also the graph of the function $f(x)=\frac{1}{x}$.

Center

The center of a rectangular hyperbola has coordinates (h, k).

Vertices and Co-vertices

The rectangular hyperbola is highly symmetric. Both its major and minor axis values are equal, so that $a=b=\sqrt{2m}$. The vertices have coordinates $\left(h+\sqrt{2m},k+\sqrt{2m}\right)$ and $\left(h-\sqrt{2m},k-\sqrt{2m}\right)$.

The co-vertices have coordinates $\left(h-\sqrt{2m},k+\sqrt{2m}\right)$ and $\left(h+\sqrt{2m},k-\sqrt{2m}\right)$.

Asymptotes

The asymptotes of a rectangular hyperbola are the x– and y-axes.

Focal Points

We can use $c^2=a^2+b^2$ as before. With $a=b=\sqrt{2m}$, we find that $c=\pm 2\sqrt{m}$. Therefore the focal points are located at $\left(h+2\sqrt{m},k+2\sqrt{m}\right)$ and $\left(h-2\sqrt{m},k-2\sqrt{m}\right)$.

Equations of Hyperbolas

In analytic geometry a hyperbola is a conic section formed by intersecting a right circular cone with a plane at an angle such that both halves of the cone are intersected. This intersection produces two separate unbounded curves that are mirror images of each other.

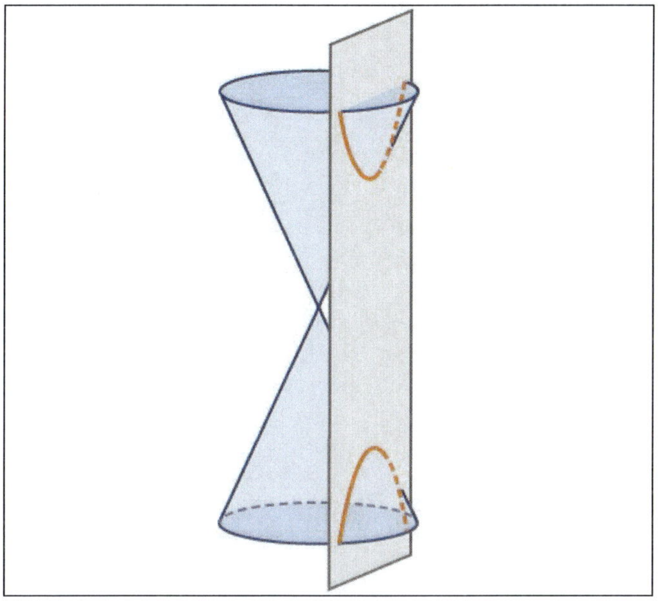

A hyperbola.

Like the ellipse, the hyperbola can also be defined as a set of points in the coordinate plane. A hyperbola is the set of all points (x, y) in a plane such that the difference of the distances between (x, y) and the foci is a positive constant.

Notice that the definition of a hyperbola is very similar to that of an ellipse. The distinction is that the hyperbola is defined in terms of the *difference* of two distances, whereas the ellipse is defined in terms of the *sum* of two distances.

As with the ellipse, every hyperbola has two axes of symmetry. The transverse axis is a line segment that passes through the center of the hyperbola and has vertices as its endpoints. The foci lie on the line that contains the transverse axis. The conjugate axis is perpendicular to the transverse axis and has the co-vertices as its endpoints. The center of a hyperbola is the midpoint of both the transverse and conjugate axes, where they intersect. Every hyperbola also has two asymptotes that pass through its center. As a hyperbola recedes from the center, its branches approach these asymptotes. The central rectangle of the hyperbola is centered at the origin with sides that pass through each vertex and co-vertex; it is a useful tool for graphing the hyperbola and its asymptotes. To sketch the asymptotes of the hyperbola, simply sketch and extend the diagonals of the central rectangle.

Hyperbolas are positioned vertically or horizontally in the coordinate plane; the axes will either lie on or be parallel to the x– and y-axes. We will consider two cases: those that are centered at the origin, and those that are centered at a point other than the origin.

Standard Form of the Equation of a Hyperbola Centered at the Origin

Let (−c, 0) and (c, 0) be the foci of a hyperbola centered at the origin. The hyperbola is the set of all points (x, y) such that the difference of the distances from (x, y) to the foci is constant.

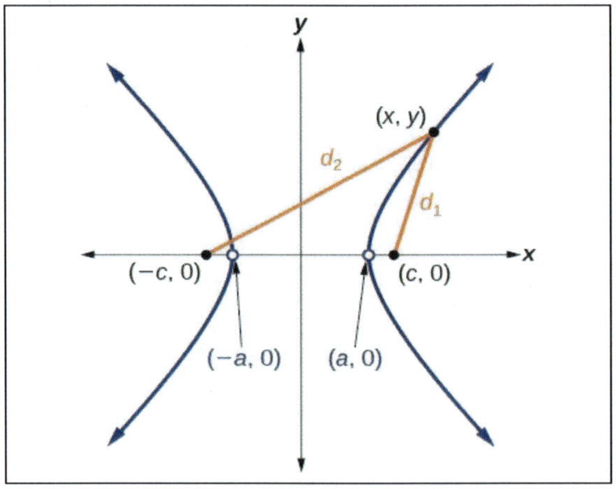

If (a, 0) is a vertex of the hyperbola, the distance from (−c, 0) to (a, 0) is a− (−c) = a + c. The distance from (c, 0) to (a, 0) is c − a. The difference of the distances from the foci to the vertex is,

$$(a+c)-(c-a)=2a$$

If (x, y) is a point on the hyperbola, we can define the following variables:

$$d_2 = \text{the distance from}(-c,0)\text{ to }(x,y)$$
$$d_1 = \text{the distance from }(c,0)\text{ to }(x,y)$$

By definition of a hyperbola, $|d_2-d_1|$ is constant for any point (x,y) on the hyperbola. We know that the difference of these distances is 2a for the vertex $(a,0)$. It follows that $|d_2-d_1|=2a$ for any point on the hyperbola. The derivation of the equation of a hyperbola is based on applying the distance formula. The standard form of an equation of a hyperbola centered at the origin with vertices $(\pm a,0)$ and co-vertices $(0\pm b)$ is $\frac{x^2}{a^2}-\frac{y^2}{b^2}=1$.

Standard Forms of the Equation of a Hyperbola with Center (0, 0)

The standard form of the equation of a hyperbola with center (0, 0) and transverse axis on the *x*-axis is,

$$\frac{x^2}{a^2}-\frac{y^2}{b^2}=1$$

where

- The length of the transverse axis is 2a.
- The coordinates of the vertices are (± a, 0).
- The length of the conjugate axis is 2b.
- The coordinates of the co-vertices are (0, ± b).

- The distance between the foci is 2c, where $c^2 = a^2 + b^2$.
- The coordinates of the foci are $(\pm c, 0)$.
- The equations of the asymptotes are $y = \pm \frac{b}{a} x$.

The standard form of the equation of a hyperbola with center (0, 0) and transverse axis on the *y*-axis is,

$$\frac{y^2}{a^2} - \frac{x^2}{b^2} = 1$$

where

- The length of the transverse axis is 2a.
- The coordinates of the vertices are $(0, \pm a)$.
- The length of the conjugate axis is 2b.
- The coordinates of the co-vertices are $(\pm b, 0)$.
- The distance between the foci is 2c, where $c^2 = a^2 + b^2$.
- The coordinates of the foci are $(0, \pm c)$.
- The equations of the asymptotes are $y = \pm \frac{a}{b} x$.

Note that the vertices, co-vertices, and foci are related by the equation $c^2 = a^2 + b^2$. When we are given the equation of a hyperbola, we can use this relationship to identify its vertices and foci.

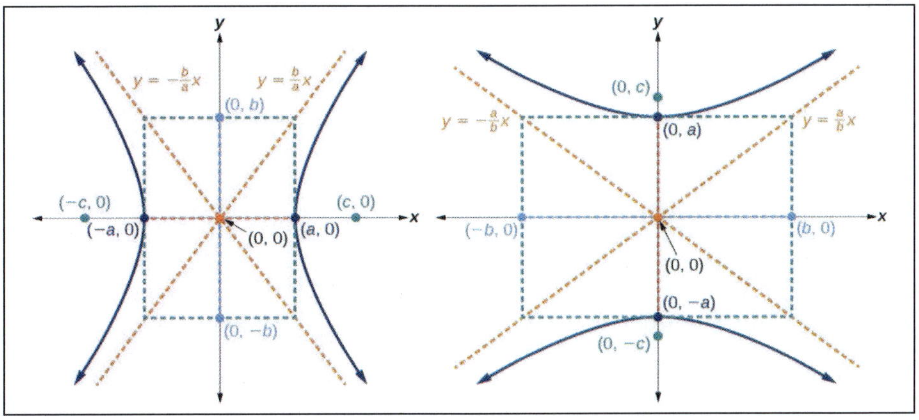

(a) Horizontal hyperbola with center (0, 0) (b) Vertical hyperbola with center (0,0).

Writing Equations of Hyperbolas in Standard Form

Just as with ellipses, writing the equation for a hyperbola in standard form allows us to calculate the key features: its center, vertices, co-vertices, foci, asymptotes, and the lengths and positions of the transverse and conjugate axes. Conversely, an equation for a hyperbola can be found given its key features. We begin by finding standard equations for hyperbolas centered at the origin. Then we will turn our attention to finding standard equations for hyperbolas centered at some point other than the origin.

Hyperbolas Centered at the Origin

Reviewing the standard forms given for hyperbolas centered at (0, 0), we see that the vertices, co-vertices, and foci are related by the equation $c^2 = a^2 + b^2$. Note that this equation can also be rewritten as $b^2 = c^2 - a^2$. This relationship is used to write the equation for a hyperbola when given the coordinates of its foci and vertices.

Hyperbolas Not Centered at the Origin

Like the graphs for other equations, the graph of a hyperbola can be translated. If a hyperbola is translated h units horizontally and k units vertically, the center of the hyperbola will be (h, k). This translation results in the standard form of the equation, with x replaced by (x−h) and y replaced by (y−k).

The asymptotes of the hyperbola coincide with the diagonals of the central rectangle. The length of the rectangle is 2a and its width is 2b. The slopes of the diagonals are $\pm\frac{b}{a}$, and each diagonal passes through the center (h, k). Using the point-slope formula, it is simple to show that the equations of the asymptotes are $y = \pm\frac{b}{a}(x-h)+k$.

The standard form of the equation of a hyperbola with center (h, k) and transverse axis parallel to the *y*-axis is,

$$\frac{(y-k)^2}{a^2} - \frac{(x-h)^2}{b^2} = 1$$

where

- The length of the transverse axis is 2a.
- The coordinates of the vertices are (h, k ± a).
- The length of the conjugate axis is 2b.
- The coordinates of the co-vertices are (h ± b, k).
- The distance between the foci is 2c, where $c^2 = a^2 + b^2$.
- The coordinates of the foci are (h, k ± c).

Using the reasoning above, the equations of the asymptotes are $y = \pm\frac{a}{b}(x-h)+k$.

Like hyperbolas centered at the origin, hyperbolas centered at a point (h, k) have vertices, co-vertices, and foci that are related by the equation $c^2 = a^2 + b^2$. We can use this relationship along with the midpoint and distance formulas to find the standard equation of a hyperbola when the vertices and foci are given.

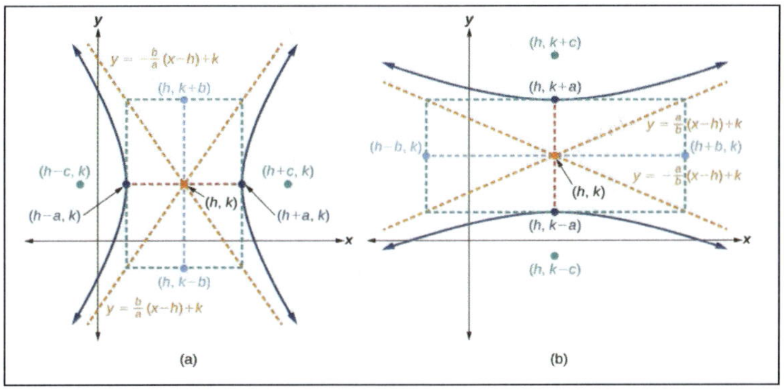

(a) Horizontal hyperbola with center (h, k) (b) Vertical hyperbola with center (h, k).

Parabola

In mathematics, the parabola is a conic section generated by the intersection of a right circular conical surface and a plane parallel to a generating straight line of that surface. A parabola can also be defined as locus of points in a plane which are equidistant from a given point (the focus) and a given line (the directrix).

A particular case arises when the plane is tangent to the conical surface. In this case, the intersection is a degenerate parabola consisting of a straight line.

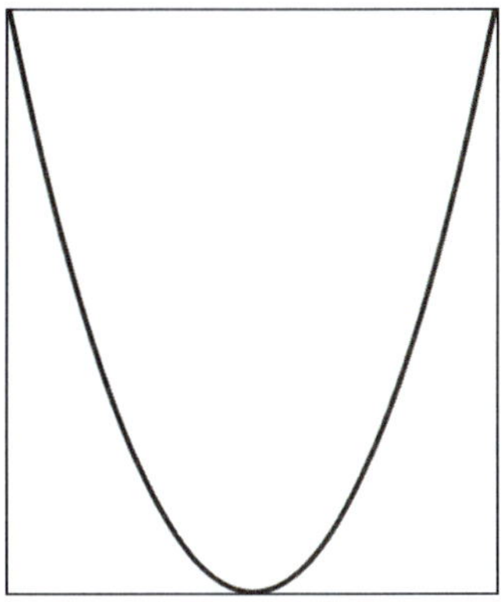
A parabola.

The parabola is an important concept in abstract mathematics, but it is also seen with considerable frequency in the physical world, and there are many practical applications for the construct in engineering, physics, and other domains.

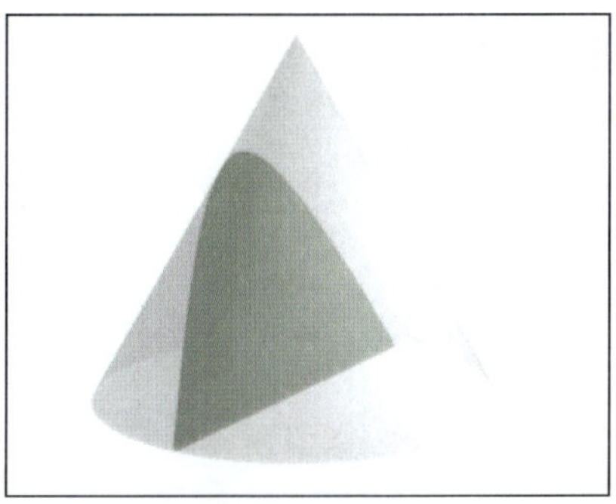
A parabola (shaded green) is a conic section.

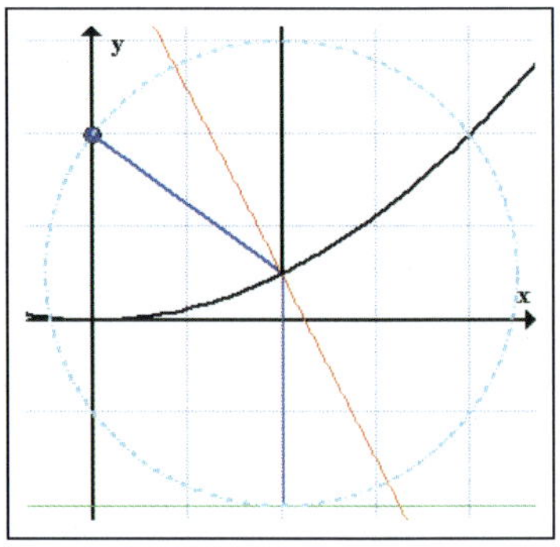

A graph showing the reflective property, the directrix (green), and the lines connecting the focus and directrix to the parabola (blue).

Analytic Geometry Equations

In Cartesian coordinates, a parabola with an axis parallel to the *y* axis with vertex (*h*, *k*), focus (*h*, *k* + *p*), and directrix *y* = *k* - *p*, with *p* being the distance from the vertex to the focus, has the equation with axis parallel to the y-axis,

$$(x-h)^2 = 4p(y-k)$$

or, alternatively with axis parallel to the x-axis,

$$(y-k) = \frac{1}{4p}(x-h)^2$$

More generally, a parabola is a curve in the Cartesian plane defined by an irreducible equation of the form,

$$Ax^2 + Bxy + Cy^2 + Dx + Ey + F = 0$$

such that $B^2 = 4AC$, where all of the coefficients are real, where $A \neq 0$ or $C \neq 0$, and where more than one solution, defining a pair of points (x, y) on the parabola, exists. That the equation is irreducible means it does not factor as a product of two not necessarily distinct linear equations.

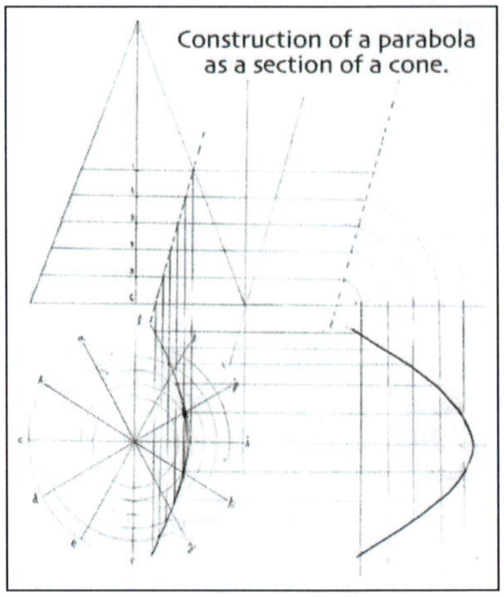

Parabolas are conic sections.

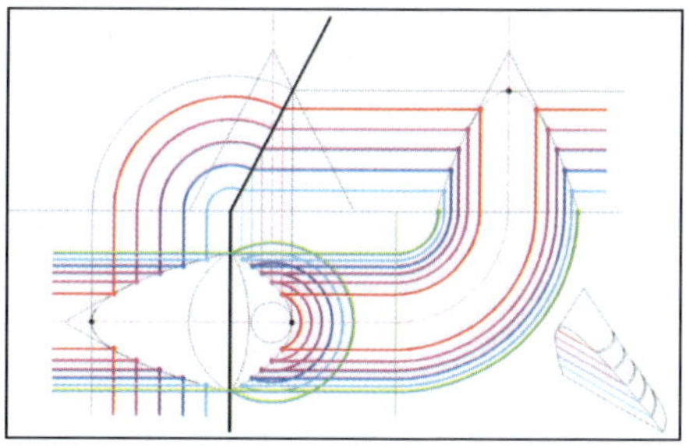

A parabola may also be characterized as a conic section with an eccentricity of 1. As a consequence of this, all parabolas are similar. A parabola can also be obtained as the limit of a sequence of ellipses where one focus is kept fixed as the other is allowed to move arbitrarily far away in one direction. In this sense, a parabola may be considered an ellipse that has one focus at infinity. The parabola is an inverse transform of a cardioid.

A parabola has a single axis of reflective symmetry, which passes through its focus and is perpendicular to its directrix. The point of intersection of this axis and the parabola is called the vertex. A parabola spun about this axis in three dimensions traces out a shape known as a paraboloid of revolution.

The parabola is found in numerous situations in the physical world.

Cartesian

Vertical Axis of Symmetry

$$(x-h)^2 = 4p(y-k)$$
$$y = a(x-h)^2 + k$$
$$y = ax^2 + bx + c$$

where $a = \dfrac{1}{4p};\ b = \dfrac{-h}{2p};\ c = \dfrac{h^2}{4p} + k;$

$$h = \frac{-b}{2a};\ k = \frac{4ac - b^2}{4a}.$$
$$x(t) = 2pt + h;\ y(t) = pt^2 + k$$

Horizontal Axis of Symmetry

$$(y-k)^2 = 4p(x-h)$$
$$x = a(y-k)^2 + h$$
$$x = ay^2 + by + c$$

where $a = \dfrac{1}{4p};\ b = \dfrac{-k}{2p};\ c = \dfrac{k^2}{4p} + h;$

$$h = \frac{4ac - b^2}{4a}; \ k = \frac{-b}{2a}.$$

$$x(t) = pt^2 + h; \ y(t) = 2pt + k$$

Semi-latus Rectum and Polar Coordinates

In polar coordinates, a parabola with the focus at the origin and the directrix on the positive x-axis, is given by the equation,

$$r(1 + \cos\theta) = l$$

where l is the *semilatus rectum*: the distance from the focus to the parabola itself, measured along a line perpendicular to the axis. Note that this is twice the distance from the focus to the apex of the parabola or the perpendicular distance from the focus to the latus rectum.

Gauss-mapped Form

A Gauss-mapped form: $(\tan^2\phi, 2\tan\phi)$ has normal $(\cos\phi, \sin\phi)$.

Derivation of the Focus

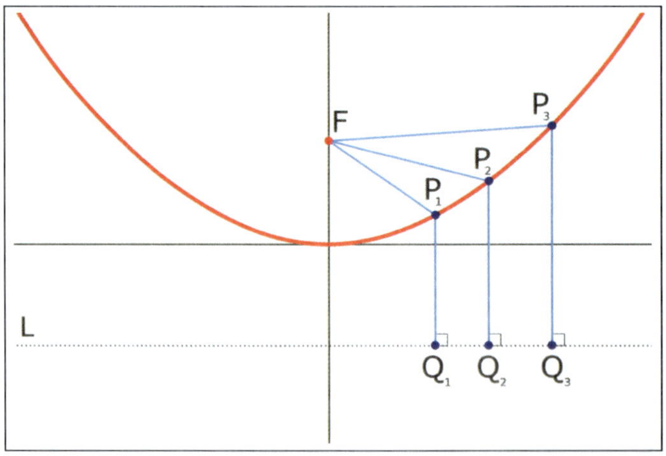

Parabolic curve showing directrix (L) and focus (F). The distance from a given point P_n to the focus is always the same as the distance from P_n to a point Q_n directly below, on the directrix.

Given a parabola parallel to the y-axis with vertex (0, 0) and with equation,

$$y = ax^2,$$

then there is a point $(0, f)$ — the focus — such that any point P on the parabola will be equidistant from both the focus and a line perpendicular to the axis of symmetry of the parabola (the *linea directrix*), in this case parallel to the x axis. Since the vertex is one of the possible points P, it follows that the linea directrix passes through the point $(0,-f)$. So for any point $P = (x, y)$, it will be equidistant from $(0, f)$ and $(x, -f)$. It is desired to find the value of f which has this property.

Let F denote the focus, and let Q denote the point at $(x,-f)$. Line FP has the same length as line QP.

$$\|FP\| = \sqrt{x^2 + (y - f)^2},$$
$$\|QP\| = y + f.$$
$$\|FP =\| = \|QP\|$$
$$\sqrt{x^2 + (ax^2 - f)^2} = ax^2 + f$$

Square both sides,

$$x^2 + (ax^2 - f)^2 = (ax^2 + f)^2$$
$$= a^2x^4 + f^2 + 2ax^2 f$$
$$x^2 + a^2x^4 + f^2 - 2ax^2 f = a^2x^4 + f^2 + 2ax^2 f$$

Cancel out terms from both sides,

$$x^2 - 2ax^2 f = 2ax^2 f,$$
$$x^2 = 4ax^2 f$$

Cancel out the x^2 from both sides (x is generally not zero),

$$1 = 4af$$
$$f = \frac{1}{4a}$$

Now let p=f and the equation for the parabola becomes,

$$x^2 = 4py$$

All this was for a parabola centered at the origin. For any generalized parabola, with its equation given in the standard form,

$$y = ax^2 + bx + c,$$

the focus is located at the point,

$$\left(\frac{-b}{2a}, \frac{-b^2}{4a} + c + \frac{1}{4a}\right)$$

and the directrix is designated by the equation,

$$y = \frac{-b^2}{4a} + c - \frac{1}{4a}$$

Reflective Property of the Tangent

The tangent of the parabola described by equation $y = ax^2$, has slope,

$$\frac{dy}{dx} = 2ax = \frac{2y}{x}$$

This line intersects the *y*-axis at the point (0,-*y*) = (0, - *a* x^2), and the *x*-axis at the point (*x*/2,0). Let this point be called *G*. Point *G* is also the midpoint of points *F* and *Q*:

$$F = (0, f),$$
$$Q = (x, -f),$$
$$\frac{F+Q}{2} = \frac{(0,f)+(x,-f)}{2} = \frac{(x,0)}{2} = (\frac{x}{2},0)$$

Since G is the midpoint of line FQ, this means that,

$$\|FG\| \cong \|GQ\|,$$

and it is already known that P is equidistant from both F and Q:

$$\|PF\| \cong \|PQ\|,$$

and, thirdly, line GP is equal to itself, therefore:

$$\Delta FGP \cong \Delta QGP$$

It follows that $\angle FPG \cong \angle GPQ$.

Line *QP* can be extended beyond *P* to some point *T*, and line *GP* can be extended beyond *P* to some point *R*. Then $\angle RPT$ and $\angle GPQ$ are vertical, so they are equal (congruent). But $\angle GPQ$ is equal to $\angle FPG$. Therefore $\angle RPT$ is equal to $\angle FPG$.

The line *RG* is tangent to the parabola at *P*, so any light beam bouncing off point *P* will behave as if line *RG* were a mirror and it were bouncing off that mirror.

Let a light beam travel down the vertical line *TP* and bounce off from *P*. The beam's angle of inclination from the mirror is $\angle RPT$, so when it bounces off, its angle of inclination must be equal to $\angle RPT$. But $\angle FPG$ has been shown to be equal to $\angle RPT$. Therefore the beam bounces off along the line *FP*: directly towards the focus.

Conclusion: Any light beam moving vertically downwards in the concavity of the parabola (parallel to the axis of symmetry) will bounce off the parabola moving directly towards the focus.

Variation of "b" in Parabolas

Vertex of a Parabola: Finding the y-coordinate

We know the x-coordinate at the vertex is $x = -\frac{b}{2a}$ so substitute it into the equation $ax^2 + bx + c$,

$$y = a\left(-\frac{b}{2a}\right)^2 + b\left(-\frac{b}{2a}\right) + c \qquad \text{Then simplify.}$$

$$= \frac{ab^2}{4a^2} - \frac{b^2}{2a} + c$$

$$= \frac{b^2}{4a} - \frac{2 \cdot b^2}{2 \cdot 2a} + c \cdot \frac{4a}{4a}$$

$$= \frac{-b^2 + 4ac}{4a}$$

$$= -\frac{b^2 - 4ac}{4a} = -\frac{D}{4a}$$

Thus, the vertex is at point,

$$\left(-\frac{b}{2a}, -\frac{D}{4a}\right)$$

Ellipse

In mathematics, an ellipse is a closed curve on a plane, such that the sum of the distances from any point on the curve to two fixed points is a constant. The two fixed points are called foci (plural of focus). A circle is a special case of an ellipse, in which the two foci coincide.

An ellipse is a type of conic section. If a conical surface is cut with a plane that does not intersect the base of the cone, the intersection of the cone and the plane is an ellipse.

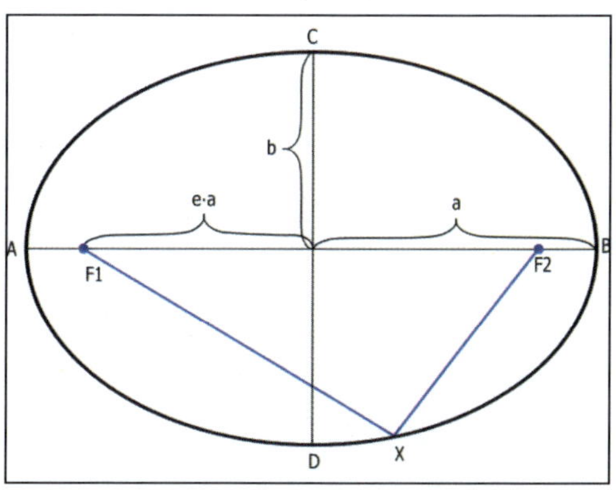

An ellipse and some of its mathematical parameters.

The orbits of planets and many comets have been found to be elliptical.

An ellipse (shaded green) is a type of conic section.

The point that lies halfway between the two foci is called the *center* of the ellipse. The line segment (or chord) AB (in the figure), which passes through the foci (F1, F2) and terminates on the ellipse, is called the *major axis*. This axis is the longest segment that can be obtained by joining two points on the ellipse. The two points at which the major axis intersects the curve are called the vertices. The line segment (chord) CD, which passes through the center, is perpendicular to the major axis, and terminates on the ellipse, is called the minor axis.

The semimajor axis (denoted by a in the figure) is one-half of the major axis and corresponds to the line segment from the center, through a focus, and to the edge of the ellipse. Likewise, the semiminor axis (denoted by b in the figure) is one-half the minor axis.

A chord drawn through a focus perpendicular to the major axis is called the latus rectum. The area enclosed by an ellipse is πab, where a and b are the semimajor and semiminor axes.

Drawing an Ellipse

An ellipse can be drawn on a sheet of paper using two pins, a pencil, and a loop of string. The pins are placed vertically at two separate points (which will become the foci of the ellipse), and the loop of string is placed around the pins and the pencil. The pencil is held vertically such that the string is taut and forms a triangle. If the pencil is moved around so that the string stays taut, it will trace out an ellipse, because the sum of the distances from the pencil to the pins will remain a constant.

Eccentricity

The shape of an ellipse can be expressed by a number called the eccentricity of the ellipse, conventionally denoted ε. It is the ratio obtained when the distance from the center to a focus is divided by the distance from the center to one of the vertices. Thus, eccentricity is a non-negative number less than one and greater than or equal to zero. An eccentricity of zero implies that the two foci occupy the same point (corresponding to the center), and in such a case, the ellipse is a circle.

For an ellipse with semimajor axis a and semiminor axis b, the eccentricity is,

$$\varepsilon = \sqrt{1 - \frac{b^2}{a^2}}.$$

The greater the eccentricity is, the larger the ratio of a to b, and therefore the more elongated the ellipse.

If c equals the distance from the center to either focus, then,

$$\varepsilon = \frac{c}{a}.$$

The distance c is known as the linear eccentricity of the ellipse. The distance between the foci is 2aε.

Equations

Algebraically, an ellipse is a curve in the Cartesian plane defined by an equation of the form,

$$Ax^2 + Bxy + Cy^2 + Dx + Ey + F =$$

such that $B^2 < 4AC$, where all of the coefficients are real, and where more than one solution, defining a pair of points (x, y) on the ellipse, exists.

Cartesian Coordinates

An ellipse with a semimajor axis a and semiminor axis b, centered at the point (h,k) and having its major axis parallel to the x-axis may be specified by the equation,

$$\frac{(x-h)^2}{a^2} + \frac{(y-k)^2}{b^2} = 1.$$

Parametric

This ellipse can be expressed parametrically as,

$$x = h + a\cos t,$$
$$y = k + b\sin t$$

where t may be restricted to the interval $-\pi \le t \le \pi$.

Polar Coordinates

If h = 0 and k = 0 (i.e., if the center is the origin (0, 0)), then we can express this ellipse in polar coordinates by the equation,

$$r = \frac{ab}{\sqrt{a^2 \sin^2 \theta + b^2 \cos^2 \theta}} = \frac{b}{\sqrt{1-\varepsilon^2 \cos^2 \theta}}$$

where ε is the eccentricity of the ellipse.

With one focus at the origin, the ellipse's polar equation is,

$$r = \frac{a\cdot(1-\varepsilon^2)}{1+\varepsilon\cdot\cos\theta}.$$

Circumference

The circumference of an ellipse is $4aE(\varepsilon)$, where the function E is the complete elliptic integral of the second kind.

Semi-latus Rectum and Polar Coordinates

The *semi-latus rectum* of an ellipse, usually denoted l (lowercase L), is the distance from a focus of the ellipse to the ellipse itself, measured along a line perpendicular to the major axis. It is related to a and b (the ellipse's semi-axes) by the formula $al = b^2$ or, if using the eccentricity, $l = a\cdot(1-\varepsilon^2)$.

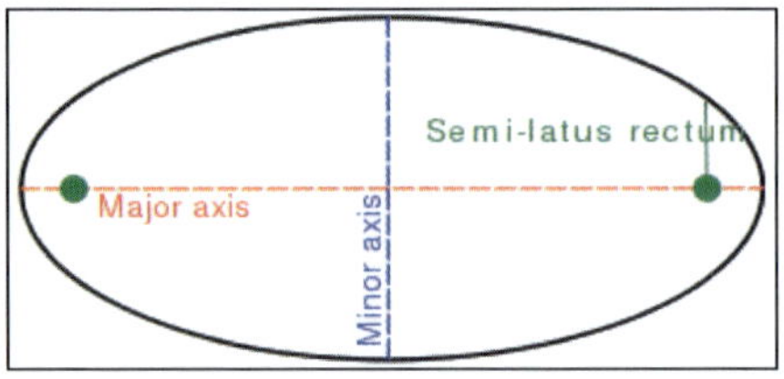

In polar coordinates, an ellipse with one focus at the origin and the other on the negative x-axis is given by the equation,

$$r\cdot(1+\varepsilon\cdot\cos\theta) = l$$

An ellipse can also be thought of as a projection of a circle: a circle on a plane at angle φ to the horizontal projected vertically onto a horizontal plane gives an ellipse of eccentricity sin φ, provided φ is not 90°.

Tangents and Normals

The tangent to the curve y = f(x) at the point (x, y) makes an angle Ψ with the positive x-axis. Then dy/dx = tan Ψ. Thus the derivative dy/dx or $f'(x)$ represents the slope of the tangent to the curve at the point (x, y).

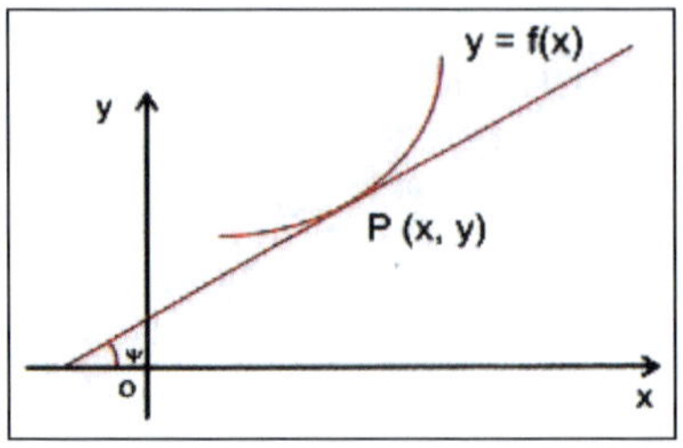

Equation of normal at (x_1, y_1),

$y - y_1 = \frac{-1}{\frac{dy}{dx}}(x - x_1)$ where the slope is calculated at the point (x_1, y_1)

In some cases, dy/dx fails to exist but even then a tangent can be drawn.

Examples:

PT is the tangent to the curve y = f(x) at the point $P(x_1, y_1)$.

PN is the normal to the curve at P.

The slope of the tangent at $P(x_1, y_1)$ is, $[dy/dx]_{(x_1, y_1)}$.

The slope of the normal at $P(x_1, y_1)$ is, $-1/[dy/dx]_{(x_1, y_1)}$.

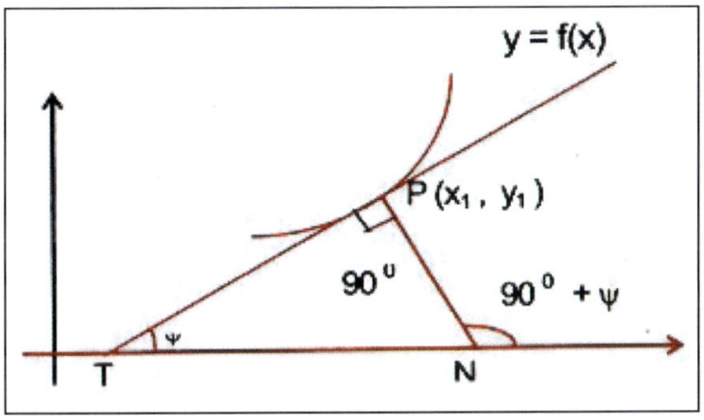

Hence the equation of the tangent PT is, $y - y_1 = [dy/dx]_{(x_1, y_1)}(x - x_1)$, and the equation of the normal PN is, $y - y_1 = -1/[dy/dx]_{(x_1, y_1)}(x - x_1)$.

If $(dy/dx)_{(x_1, y_1)} = 0$, then the equation of the normal would be $x = x_1$.

If the equation of the curve is in the parametric form x = f(t) and y = g(t),

Then dy/dx = dy/dt/dx/dt = (g' (t))/(f' (t)).

Thus the equations of the tangent and the normal are:

y - g(t) = (g' (t))/(f' (t)) (x - f(t))

and f'(t)[x - f(t)] + g'(t)[y - g(t)] = 0.

Some important points:

- If dy/dx at $(x_1, y_1) = 0$ then the tangent is parallel to x-axis and conversely.
- If the tangent is parallel to ax + by + c = 0 then dy/dx = -a/b.
- If dy/dx at $(x_1, y_1) \rightarrow \infty$ i.e. dx/dy at $(x_1, y_1) = 0$ then the tangent is perpendicular to x- axis.
- If the tangent at $P(x_1, y_1)$ is equally inclined to the coordinate axes, then dy/dx at $(x_1, y_1) = \pm 1$.

- If the tangent makes equal non-zero intercepts on the coordinate axis, then dy/dx at $(x_1, y_1) = -1$.
- If the tangent cuts off from the coordinate axis equal distance from the origin then dy/dx = ±1.

Some common parametric coordinates on a curve:

- For $x^{2/3} + y^{2/3} = a^{2/3}$, assume the parametric coordinates $x = a\cos^3\theta$ and $y = a\sin^3\theta$.
- For $\sqrt{x} + \sqrt{y} = \sqrt{a}$, assume $x = \cos^4\theta$ and $y = a\sin^4\theta$.
- For $x^n / a^n + y^n / b^n = 1$, assume $x = a(\sin\theta)^{2/n}$ and $y = b(\sin\theta)^{2/n}$.
- For $y^2 = x^3$, take $x = t^2$ and $y = t^3$.

Angle of Intersection of Two Curves

The angle of intersection of two curves at a point P is defined as the angle between the two tangents to the curve at their point of intersection.

Let there be two curves $y = f_1(x)$ and $y = f_2(x)$ which intersect each other at point (x_1, y_1). If we draw tangents to these curves at the intersecting point, the angle between these tangents is called the angle between two curves.

Let $m_1 = (df_1(x))/dx \,\big|_{(x = x_1)}$ and $m_2 = (df_2(x))/dx \,\big|_{(x = x_1)}$

And both m_1 and m_2 are finite.

The acute angle between the curves is given by,

$$\theta = \tan^{-1}\left|(m_1 - m_2)/(1 + m_1 m_2)\right|$$

Note: (p - q) is also an angle between lines.

If m_1 (or m_2) is infinity the angle is given by $\theta = |\pi/2 - \theta_1|$ where,

$$\theta_1 = \tan^{-1} m_2 \left(or \ \tan^{-1} m_1\right)$$

In the figure given below, Φ is the angle between the two curves, which is given by,

$$\Phi = \Psi_1 - \Psi_2$$

$$\Rightarrow \tan\Phi = \tan(\Psi_1 - \Psi_2)$$

$$= (\tan\Psi_1 - \tan\Psi_2)/(1 + \tan\Psi_1 \tan\Psi_2), \text{ where } \tan\Psi_1 = f'(x_1) \text{ and } \tan\Psi_2 = g'(x_1).$$

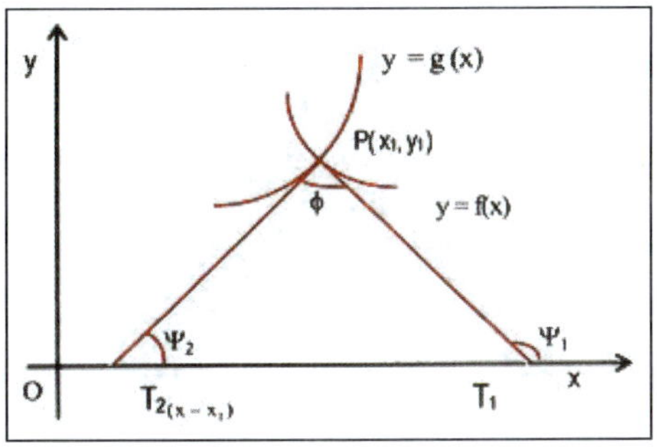

Two curves are said to cut each other orthogonally if the angle between them is a right angle, in which case we will have,

$$\tan\Psi_1 \tan\Psi_2 = -1.$$

Two curves touch each other if the angle between the tangents to the curves at the point of intersection is 0°, in which case we will have,

$$\tan \Psi_1 = \tan \Psi_2.$$

Illustration: Show that the curves $ax^2 + by^2 = 1$ and $cx^2 + dy^2 = 1$ cut each other orthogonally if, $1/a - 1/b = 1/c - 1/d$.

Solution: Let the two curves cut each other at the point (x_1, y_1). Then,

$$ax_1^2 + by_1^2 = 1$$

and $cx_1^2 + dy_1^2 = 1.$

From $ax_1^2 + by_1^2 = 1$ and $cx_1^2 + dy_1^2 = 1.$, we get

$$(a-c)x_1^2 + (b-d)y_1^2 = 0.$$

Slope of the tangent to the curve $ax^2 + by^2 = 1$, at (x_1, y_1) is given by,

$$\tan \Psi_1 = [dy/dx]_{(x_1,y_1)} = -ax_1/by_1.$$

Slope of the tangent to the curve $cx^2 + dy^2 = 1$ at (x_1, y_1) is given by,

$\tan \Psi_2 = [dy/dx]_{(x1,y1)} = -cx_1/dy_1$. If the two curves cut orthogonally, we must have,

$$(-ax_1/by_1)(-cx_1/dy_1) = -1 \Rightarrow acx_1^2 + bdy_1^2 = 0.$$

From $(a-c)x_1^2 + (b-d)y_1^2 = 0.$ and $(-ax_1/by_1)(-cx_1/dy_1) = -1 \Rightarrow acx_1^2 + bdy_1^2 = 0.$ we have $(a-c)/ac = (b-d)/bd \Rightarrow 1/a - 1/c = 1/c - 1/d.$

Illustration: Prove that the tangent lines to the curve $y^2 = 4ax$ at points where $x = a$ are at right angles to each other.

Solution:

$y^2 = 4ax$

When x = a

$y^2 = 4a^2$

$\Rightarrow y = \pm 2a$

Points are (a, 2a) and (a, - 2a)

Slope of the tangent of the curve $y^2 = 4ax$ is

$y\, dy/dx = 4a$

$dy/dx = 2a/y$

for y = 2a, $dy/dx = 1 = m_1$

for y = -2a $dy/dx = -1 = m_2$

$m_1 m_2 = -1$

Length of the Tangent, Normal, Subtangent and Subnormal

Let P(x, y) be any point on y = f(x). Let the tangent drawn at 'P' meet the x-axis at 'T', and normal drawn at 'P' meets the x-axis at 'N'. PT is called the length of the tangent and PN is called the length of the normal.

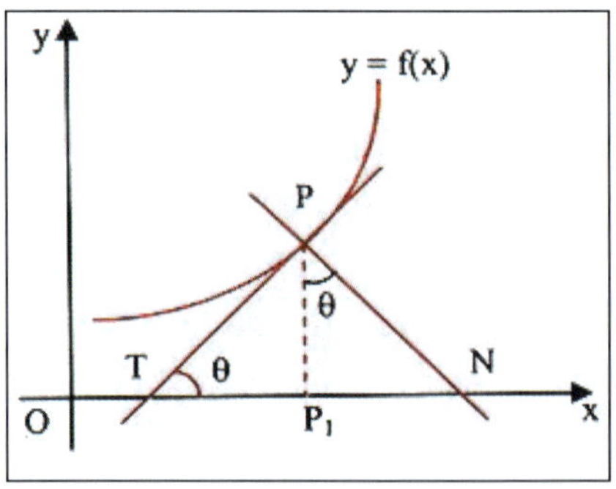

If 'P_1' be the projection of the point P on the x-axis then TP_1 is called the sub-tangent (projection of line segment PT on the x-axis) and NP_1 is called the sub normal (projection of line segment PN on the x-axis).

Tangent:

$$TP = PP_1 \operatorname{cosec} \theta = y \sqrt{\left(1+\cot^2\theta\right)}$$

$$\left|\frac{y\sqrt{1+(dy/dx)^2}}{dy/dx}\right|$$

Subtangent:

$TP_1 = PP_1 \cot \theta$

$$= \left| \frac{y}{dy/dx} \right|$$

Normal:

$PN = PP_1 \sec \theta = y \sqrt{(1 + \tan^2\theta)}$

$$= \left| y\sqrt{1 + (dy/dx)^2} \right|$$

Subnormal:

$NP_1 = PP_1 \tan y = |y(dy/dx)|$

Interpretation of dy/dt as a Rate Measure

We know that by the derivative dr/dt, we mean the rate of change of distance r with respect to time t. In the similar way, when some quantity y varies or changes with some other quantity x while satisfying some rule y = f(x), then the derivative dy/dx i.e. f'(x) implies the rate of change of y with respect to x.

References

- Analytic-geometry, science: britannica.com, Retrieved 08 January, 2019
- Coordinate-geometry, guides-maths: toppr.com, Retrieved 19 February, 2019
- Cartesian-coordinate-system: newworldencyclopedia.org, Retrieved 27 July, 2019
- Spherical-coordinate-system, math-for-physics - 103529158: fiberoptics4sale.com, Retrieved 13 August, 2019
- Conic-section: newworldencyclopedia.org, Retrieved 12 March, 2019
- The-hyperbola, boundless-algebra: courses.lumenlearning.com, Retrieved 11 July, 2019
- Parabola: newworldencyclopedia.org, Retrieved 19 April, 2019
- Tangents-and-normal, differential-calculus: askiitians.com, Retrieved 16 March, 2019
- Ellipse: newworldencyclopedia.org, Retrieved 28 August, 2019

Permissions

All chapters in this book are published with permission under the Creative Commons Attribution Share Alike License or equivalent. Every chapter published in this book has been scrutinized by our experts. Their significance has been extensively debated. The topics covered herein carry significant information for a comprehensive understanding. They may even be implemented as practical applications or may be referred to as a beginning point for further studies.

We would like to thank the editorial team for lending their expertise to make the book truly unique. They have played a crucial role in the development of this book. Without their invaluable contributions this book wouldn't have been possible. They have made vital efforts to compile up to date information on the varied aspects of this subject to make this book a valuable addition to the collection of many professionals and students.

This book was conceptualized with the vision of imparting up-to-date and integrated information in this field. To ensure the same, a matchless editorial board was set up. Every individual on the board went through rigorous rounds of assessment to prove their worth. After which they invested a large part of their time researching and compiling the most relevant data for our readers.

The editorial board has been involved in producing this book since its inception. They have spent rigorous hours researching and exploring the diverse topics which have resulted in the successful publishing of this book. They have passed on their knowledge of decades through this book. To expedite this challenging task, the publisher supported the team at every step. A small team of assistant editors was also appointed to further simplify the editing procedure and attain best results for the readers.

Apart from the editorial board, the designing team has also invested a significant amount of their time in understanding the subject and creating the most relevant covers. They scrutinized every image to scout for the most suitable representation of the subject and create an appropriate cover for the book.

The publishing team has been an ardent support to the editorial, designing and production team. Their endless efforts to recruit the best for this project, has resulted in the accomplishment of this book. They are a veteran in the field of academics and their pool of knowledge is as vast as their experience in printing. Their expertise and guidance has proved useful at every step. Their uncompromising quality standards have made this book an exceptional effort. Their encouragement from time to time has been an inspiration for everyone.

The publisher and the editorial board hope that this book will prove to be a valuable piece of knowledge for students, practitioners and scholars across the globe.

Index